Human-Machine Interface Design for Process Control

Human-Machine Interface Design for Process Control

Jean-Yves Fiset

Human-Machine Interface Design for Process Control

Copyright © 2009 by ISA—Instrumentation, Systems, and Automation Society

67 Alexander Drive
P.O. Box 12277
Research Triangle Park, NC 27709

All Rights Reserved.
Printed in the United States of America.
10 9 8 7 6 5 4 3 2

ISBN: 978-1-934394-35-9

No part of this work may be reproduced, stored in a retrieval system, or transmitted in any form or by any means, electronic, mechanical, photocopying, recording or otherwise, without the prior written permission of the publisher.

Notice
The information presented in this publication is for the general education of the reader. Because neither the author nor the publisher have any control over the use of the information by the reader, both the author and the publisher disclaim any and all liability of any kind arising out of such use. The reader is expected to exercise sound professional judgment in using any of the information presented in a particular application. Additionally, neither the author nor the publisher have investigated or considered the affect of any patents on the ability of the reader to use any of the information in a particular application. The reader is responsible for reviewing any possible patents that may affect any particular use of the information presented. Any references to commercial products in the work are cited as examples only. Neither the author nor the publisher endorses any referenced commercial product. Any trademarks or trade names referenced belong to the respective owner of the mark or name. Neither the author nor the publisher makes any representation regarding the availability of any referenced commercial product at any time. The manufacturer's instructions on use of any commercial product must be followed at all times, even if in conflict with the information in this publication.

Library of Congress Cataloging-in-Publication Data

Fiset, Jean-Yves.
  Human-machine interface design for process control / Jean-Yves Fiset.
    p. cm.
  Includes bibliographical references and index.
  ISBN 978-1-934394-35-9
  1. Process control--Data processing. 2. Human-machine systems. I. Title.
  TS156.8.F565 2009
  670.42'75--dc22
                                            2008037769

To Cécile, Elisabeth, and Monique, for the kindness you bring to this world. To mentors, friends and colleagues, whose questions and challenges have helped clarify my thinking and understanding.

—Jean-Yves Fiset, Eng., Ph.D.

Table of Contents

Preface xi
About the Author xiii
List of Figures xv
List of Tables xvii

1 Introduction 1
Audience 2
Why Design an HMI? 2
Organization of the Book 5
Paper-based Publications 6

2 Designing a New HMI 7
Understanding the User 7
  Visual Perception 8
  Memory 10
Decision-making 11
  Internal or Mental Model 12
  Human Error 13
  Operator Monitoring 15
Linking HMI Design and System Design 16
Standards and Guidelines 17
Overall HMI Design Process 19
Detailed Description of the HMI Design Method 22

Example: Newsprint Papermaking   22
Planning the User-Centered Design Process   24
   Description of the Planning Phase   24
   Example: Carrying Out the Planning Phase   25
Understanding and Specifying the Context of Use   26
   Description of the Context of Use   27
   Example: Analyzing the Context of Use   30
Specifying User and Organizational Requirements   34
   Description of User and Organizational Requirements   34
   Example: Analyzing User and Organizational Requirements   36
Documentation   38
Producing Design Solutions   38
   Production of Design Solutions   40
   Example: Producing Design Solutions   49
Specification of the Resulting HMI   52
Integrating Software-based and Hardwired HMIs   53
Transition   54
References and Additional Readings   54
Paper-based Publications   54
Web-based Resources   56

3   Evaluating the Design Against the Requirements 57
Description of the Evaluation of Design Solutions   57
   Heuristic Evaluations   57
   Example of Heuristic Evaluation   61
   Usability Testing   63
   Usability Testing Example   67
Transition   71
References and Additional Readings   71
Paper-based Publications   71
Web-based Resources   72

4   Specifying an HMI 73
Standards and Guidelines   74
How to Specify an HMI   75
   Ground Rules   75

Starting to Specify: Introduction   77
Specifying Common Elements: General Section   77
Specifying Individual Components: Detailed Specification   78
An Example of Specification   79
References and Additional Readings   82
Paper-based Publications   82
Web-based Resources   82

5 Improving an Existing HMI 83
Why Improve an Existing HMI?   83
Improving an Existing HMI—With Design Basis   85
Improving an Existing HMI—Without Design Basis   89
Transition   92
References and Additional Readings   93

6 Continuous, Batch, Discrete and Hybrid Applications 95
Designing HMIs for Different Types of Applications   95
Transition   98

7 Integrating Heterogeneous HMIs 99
What Integration Means   99
Simple Integration Strategies   100
Component-specific Integration Rules   102
Allocating Control and Monitoring Functionality to Computerized or Hardwired Panels   102
Using Large Screen Displays   105
Using Portable and Wireless Devices   107
Combining HMI Resources   109
Transition   109
References and Additional Readings   109
Paper-based Publications   110
Web-based Resources   110

8 Overall Organization of HMIs in a Control Room 111
Workstations   112
Advantages, Disadvantages and Utilization   112

Organizing Principles—Workstations 114
Organizing Principles—Additional HMI Components 117
Locating the HMI Components 118
Transition 122
References and Additional Readings 122
Paper-based Publications 122
Web-based Resources 124

9 Additional HMI Components 125
Operating Procedures 125
    The Benefits of Procedures 126
    The Development Process 127
        Define the Purpose and Scope of the Procedures 127
        Defining Training Requirements for Using the Procedures 128
        Specifying How to Write, Validate and Verify Procedures 128
        Determining the Content of Procedures 130
    Procedure Design Principles and Rules 130
        Structuring with Goals 131
        Complying with and Deviating from Procedure Steps 131
        Writing Procedure Steps 132
        Considerations for Computerized Procedures 132
    Integration with the HMI 134
        Decision-Support Systems 136
        Challenges 137
    Integration into the HMI 138
    Conclusion 139
    Paper-based Publications 140

A Design Guidelines and Components 141
Computer-based HMIs 141
Hardwired panels 142
Large screen displays (LSDs) 142

Glossary 165
Index 167

Preface

This book was written to share a passion: designing human-machine interfaces that enable organizations to be more efficient and safer, and users to be more satisfied about what they do.

When the author was given his first design assignment for a new human-machine interface, he realized that in spite of the academic training he had received in this area, he was still ill-equipped to produce the required design. After several years of experience and countless projects, as well as teaching numerous courses on the same topic, the author realized that it takes some key ingredients to produce good design:

- A good understanding of how users comprehend a system and control it. In some cases, this understanding will even force the designer to let go of preconceived ideas about how operators work (the author had to, anyway).

- A wide knowledge of the appropriate design principles and rules. Those will be both generic (applicable across domains) and specific (for a given domain, such as industrial process control). The design principles and rules are usually drawn from specific standards, guidelines and even textbooks. We do not intend to repeat all of those in this book, but rather to provide you with pointers to the appropriate sources of information.

- A strong understanding of engineering design in general and of the processes used for designing human-machine interfaces in particular. It is necessary to both understand how to do the interface design itself, and, sometimes just as importantly, to understand where and how the interface design will impact the overall design project.

There are also unexpected benefits in understanding human-machine interface design. As you will discover, most equipment or even pieces of information (e.g., documents and Web pages) has an interface. Interestingly, the approach used to design those interfaces is the same as the one described in this book; usually, the differences reside in the detailed design rules, while the overall process and principles remain the same. In other words, the material in this book will be useful even outside the realm of industrial process control (in much the same way as arithmetic is used anywhere calculations are required).

You will hopefully also find that human-machine interface design offers a unique occasion to positively affect human and organizational performance, and this is often an exhilirating experience.

Finally, I hope that you will enjoy this material as much as I have enjoyed writing about it.

—Jean-Yves Fiset, Eng., Ph.D.

About the Author

Jean-Yves Fiset started his career as an electronic technologist for a process automation firm where he acquired a fair bit of practical experience on installation and troubleshooting, and explaining how those systems work. This is where he realized the importance of the interface between the users and the equipment. He then completed a bachelor of engineering (industrial engineering) and a Master of Applied Sciences (industrial engineering) specializing in human factors and artificial intelligence at École Polytechnique de Montréal. After graduating, he went to work for Atomic Energy of Canada (AECL) as a research scientist specializing in human-machine interfaces, automated emergency operating procedures and control rooms. Concurrently, he started a Ph.D., again at École Polytechnique de Montréal, where his dissertation dealt with the design of human-machine interfaces for monitoring and controlling complex systems. After leaving AECL, Dr. Fiset established Systèmes Humains-Machines Inc. ("Shumac"), which specializes in the analysis, design and validation of high-performance human-machine interfaces in the aerospace, nuclear, military, Web, industrial, banking and software fields. He also teaches (human factors, cognitive engineering, artificial intelligence) at the undergraduate and graduate levels since 1991. Overall, he has been involved with using and (sometimes!) understanding complex systems for nearly 30 years.

List of Figures

Figure 1–1	Growth of Bandwidth Offered by HMIs	4
Figure 2–1	Determination of Character Height	9
Figure 2–2	HMI Design and the Overall Design Cycle	16
Figure 2–3	Overall HMI Design Process	20
Figure 2–4	Fictitious Newsprint Machine	23
Figure 2–5	Fragment of a Model of the Operator's Task	33
Figure 2–6	Example of Transition from HMI Architecture to Detailed Design	47
Figure 2–7	Example of Reduction in Perceived Complexity	47
Figure 2–8	Example of Using Human Pattern Recognition	48
Figure 2–9	Example Template for Display	50
Figure 2–10	Sample Monitoring Display and HTA Fragment	51
Figure 3–1	Sample Window for Heuristic Evaluation	62
Figure 3–2	Sample Hardwired Panel	68
Figure 3–3	Mock-up 1	70
Figure 3–4	Mock-up 1 with Indication of Results	70
Figure 3–5	Mock-up 2	71

Figure 4–1	Format for Individual Window or Category of Windows	79
Figure 5–1	Results Gap	85
Figure 5–2	Process to Enhance an Existing HMI—with Design Basis	87
Figure 5–3	Process to Enhance an Existing HMI—without Design Basis	90
Figure 6–1	Impacts of Application-specific Design Principles on the HMI Design Process	98
Figure 7–1	Example of a Heterogeneous Environment	100
Figure 7–2	Information Integration	105
Figure 8–1	(a) Side-by-Side Arrangement. (b) Alternate Arrangement.	117
Figure 8–2	Raw Data—Moves	121
Figure 8–3	Summary of Moves—Bilateral	121
Figure 8–4	Results from Analysis and Optimization	121

List of Tables

Table 2–1	Standards and Guidelines Relevant to HMI Design	18
Table 2–2	HMI Design Project Schedule	27
Table 2–3	Examples of User Characteristics	29
Table 2–4	Key Characteristics of Generic Tasks	43
Table 3–1	List of Heuristics	60
Table 4–1	Format for Individual Window or Category of Windows (Reference Figure 4–1)	80
Table 4–2	Example Table of Contents	81
Table 7–1	Comparison of LSD Technologies	106
Table 8–1	General Characteristics of Workstations	113
Table A–1	Computer-based HMI	143
Table A–2	Windows	147
Table A–3	Display Components	150
Table A–4	Inputs	154
Table A–5	Hardwired Panels—General	158
Table A–6	Hardwired Panels—Controls	159
Table A–7	Hardware Panels—Displays	161
Table A–8	Large Screen Displays	162
Table A–9	Procedures	163

1

Introduction

This book aims to provide users and designers of industrial control and monitoring systems with an easy-to-use—yet effective—method to configure, design, and validate human-machine interfaces (HMIs). Such systems include distributed control systems (DCSs); supervisory control and data acquisition systems (SCADAs); and stand-alone units.

- **Distributed Control Systems (DCSs)** are typically real-time, fault-tolerant systems for continuous and complex batch process applications. DCSs were developed initially for continuous flow processes that required loop, analog, and limited discrete control. A DCS, while functionally integrated, consists of sub-systems that may be physically separate and remotely located from one another.

- **Supervisory Control and Data Acquisition System (SCADA)** is typically a generic name for a computerized system capable of gathering and processing data and applying operational controls over long distances, such as is used with power transmission and distribution and pipeline systems. SCADA systems are designed for unique communication challenges (delays, data integrity, etc.) resulting from the various media that must be used, such as phone lines, microwave, satellite, and so on. SCADA systems are usually shared, rather than dedicated.

■ **Stand-alone Units** are typically simple, embedded systems that perform pre-defined tasks, usually with very specific requirements.

This book discusses the overall HMI design process; how that process relates to system design; detailed design methods, principles, and rules for individual displays and groups of displays; and integrating both software-based and hardwired HMIs.

It also provides guidance on the design of HMIs for other, less common, yet important components, such as expert systems and electronically displayed operating procedures. With the information contained in this book, a user or designer can determine how to configure or design a whole new set of displays for a system, or how to enhance specific elements of an existing or planned HMI.

The material originates, to a large extent, from a graduate course (IND 6408 *Human Factors in Process Control*) and a fourth-year course (IND 4803 *Cognitive Engineering for Complex Systems*) that the author has been fortunate to teach for a number of years at the Department of Mathematics and Industrial Engineering of École Polytechnique de Montréal. A substantial proportion of the material is also drawn from the professional practice of the author in HMI analysis, design, and validation in domains ranging from nuclear power plants to the Web.

Audience

The primary audience intended for this book is designers and developers of HMIs; however, it will also assist project leaders and managers, engineers, and system integrators in ensuring that the systems they design will work synergistically with operators, thus leading to improved safety and productivity.

The book was also written with operators in mind. Having been in field service for a number of years, the author recognizes that operators and their supervisors optimize processes on a day-to-day basis. It is hoped this book will help them better understand how to enhance and optimize existing HMIs, and to express their operational requirements in a way that can be more easily understood by HMI designers.

Why Design an HMI?

Having said all that, it is perhaps a good idea to explain why HMI design matters. Borrowing from a popular song[1], the HMI is "not just a pretty face" that will make the system look good. In fact, if the HMI looks aesthetically pleasing, the look will result more from an elegant design than from a conscious effort. Of course, if we can, we will always strive to make an HMI look good. However, the most important objective is to achieve usability, which is defined as:

> "... the extent to which a product can be used by specified users to achieve specified goals with effectiveness, efficiency and satisfaction in a specified context of use...."

Right away, we realize that usability has little, if anything, to do with individual preferences. Rather, a system will be termed *usable* (rather than *user-friendly*, which means precious little because it is unmeasurable) when it helps specified users to achieve specified goals.

A question frequently raised is the actual need for designing HMIs. After all, have we not designed HMIs for the last several decades? Don't operators know what they want? Doesn't management know what the operators need? The unfortunate answer to those questions is a resounding "no." Rather than taking this answer at face value, let's look at a few facts.

Figure 1–1 shows that the bandwidth made available by modern HMI hardware and software has grown exponentially over the last few decades (Cochran, 1992). Note that current Internet and Web technologies are not addressed in this figure.

Unfortunately, human capability to absorb and understand the information has not grown at the same rate as HMI bandwidth. Neither has our ability to provide the information in a comprehensible way. For this reason alone, it is justifiable to try to design HMIs that better support the operator.

An example can be taken from the petrochemical industry. In the United States alone, it has been estimated that inadequacies in the means available to deal with abnormal situations (including HMIs

---

1. Shania Twain's *Not Just a Pretty Face.*

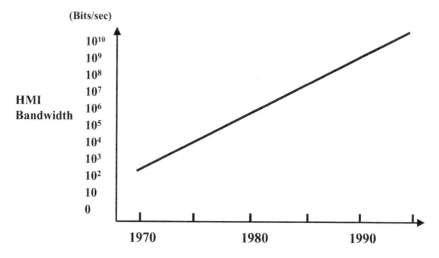

Figure 1-1 *Growth of Bandwidth Offered by HMIs*

used to identify, diagnose, and deal with those situations) cost the industry between ten and twenty billion dollars each year. Clearly, there is a need to do better and, as we will see, it is possible, and even easy, to achieve that objective.

We do not pretend that HMI design is a panacea; rather, we see it as a key component of sensible system design. Even more, given that a system's architecture is driven by the design of its interfaces (Meier & Rechtin, 2002), and since the HMI is one of the major interfaces of an interactive system, it becomes clear that the HMI's design will exert enormous influence on the architecture and design of the larger system.

This viewpoint may come as a shock to design engineers who typically look first at functionality, flow diagrams, electrical diagrams, etc. and only later worry about how the system will be used. The trick here is to realize that the system is there to support its users in achieving a set of operating goals that the organization desires to attain. One of those goals, at least for commercial organizations, is to achieve a near-optimal productivity. This goal rests, in part, on sub-goals the operators achieve through their own goals and sub-goals. Those operators' sub-goals, and the strategies they use to reach them, drive (to a large extent) HMI design and define sub-functionalities the system will have to provide. The same argument can be made for goals related to safety.

Organization of the Book

Since this book is intended for a wide and diverse audience, the material has been organized to meet various objectives. It should be noted that no assumptions have been made about the technology that will be used. While specific examples are given in the context of a given technology (e.g., software, hardwired panels), the content of the book applies to HMIs in general, ranging from individual control panels all the way to the display suite of a complete monitoring and control system, whether implemented in hardware, software, or a combination of both.

The first chapter deals with the toughest scenario: designing an HMI from scratch. Depending on the reader's role, this chapter may be read in detail (e.g., by design staff) or at a higher level, concentrating more on the process used than on the details (e.g., by operation or management staff). It explains each of the steps of the design process and offers a fictitious, yet realistic, example of what is involved.

The next chapter deals with the evaluation of the HMI that has been designed. Two important methods are introduced and discussed in detail. This chapter should be of interest to both technical and non-technical staff.

The results of the design exercise will then feed into the system requirement specification. The third chapter explains how to specify an HMI design in such a way that it can easily be integrated into the project's technical documentation.

Once an HMI and its associated systems have been designed, validated, and implemented, it will evolve to fit the needs of the organization. The fourth chapter deals with the continuous improvement of the HMI and provides methods to ensure that initial benefits, which are expected to be substantial, do not eventually erode.

Chapter five describes specific aspects of HMI design for DCS and programmable logic control (PLC) applications, while chapter six discusses the integration of heterogeneous components, such as wireless devices and third-party HMI components.

Chapter seven discusses how consoles that will host the HMI will be integrated into the control center, thus leading to a synergistic interaction between those components.

Strictly speaking, an HMI comprises all of the elements that a user will touch, see or otherwise use to carry out his or her tasks. This implies that components that have not traditionally been considered as being part of the HMI need to be considered in the design to ensure that the resulting product will optimally support the user, who is striving to achieve the organization's operational and safety goals. Chapter eight thus discusses designing and integrating decision-support systems and operating procedures into HMIs.

Rather than cluttering the book's text with endless individual design rules, those rules have been assembled into Appendices. They can be readily referenced, as required, into examples provided in the main chapters. They are also easy to find at design time.

Paper-based Publications

Cochran, Edward L., *Control Room User Interface Technology for the year 2000: Evolution or Revolution?* in Proceedings of the Human Factors Society 36th Annual Meeting 1992. pp. 460-464.

Meier, M., Rechtin, E., *The Art of Systems Architecting*, ISBN: 0-8493-0440-7 (Hardbound), CRC Press, 2002.

2

Designing a New HMI

In this chapter, we will learn how to design an HMI from scratch. A subsequent chapter will cover how to enhance or evolve an existing HMI. We are going to approach the design of a new HMI in a simple, yet systematic, fashion, using both existing standards and state-of-the-art knowledge about HMI design. To help ensure that the theory translates well into practice, we will use a small example as a vehicle to better communicate the design principles and rules.

A point worth mentioning now is that we will leverage experience acquired in several domains, ranging from biomedical devices to operator consoles and control centers found in nuclear power plants, as well as in designing for the ubiquitous World Wide Web. What benefits can the latter provide? The answer is simple: as the Web is huge, approaches and methods, imperfect as they are, have been developed to allow users to better navigate very large Web sites. Useful lessons can therefore be drawn to help the operator find and act upon the information he or she needs to contribute toward achieving operational goals. It has been found that experience in one domain can often be applied to another domain.

Understanding the User

To design successful HMIs, it is necessary to know and understand a suitable design process. However, it is also invaluable to understand how the HMI's users perceive and remember things, and how they make decisions. It is also important to realize how well they understand the process and the systems they use, and to be able to identify beforehand the types of mistakes they are likely to make.

Finally, having a good understanding of operator monitoring activities will be useful in designing interactive monitoring functions. Since there are numerous and excellent textbooks on many of these topics, we will limit our discussion to fundamental aspects that have to be understood to successfully design HMIs. We will discuss topics in more or less the order they occur during human information processing.

Visual Perception

Visual perception here refers to how users see and acquire information relevant to what they are trying to achieve. Since countless talented authors have previously covered this topic, let's concern ourselves with just key facts about how people acquire information visually.

One of the first things to realize is that the legibility of visual information is influenced by a large number of factors such as contrast, font stroke width, ambient lighting, etc. From a design point of view, however, character height has been used as the main means to affect legibility in HMIs. Fortunately, this issue has been researched in great depth, and a suitable character height can be determined by considering the parameters shown in Figure 2–1.

A suitable character height can be calculated as an angle expressed in minutes of arc. The required angle ranges from 16 to 22 minutes of arc for applications such as process control. The formula to use is:

$$H = \frac{A \times D}{3438}$$

Where A is the angle, while H and D are the character height and distance between the operator's eye and the display, respectively. For example, for a reading distance of 75 cm (typical of table-top display reading) and a character height of 20 minutes, the resulting actual character height in engineering units as shown in Figure 2–1 is:

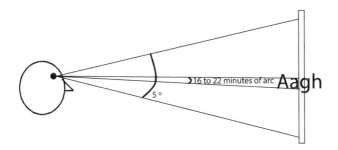

A(minutes) = ((3438 x H) / D), with H and D expressed in the same unit

Figure 2–1 *Determination of Character Height*

$$H = \frac{A \times D}{3438} = \frac{20\min \times 75\text{cm}}{3438} = 0.44\text{cm or } 4.4\text{mm}$$

The formula works fine as long as H and D are expressed in the same unit, be they centimeters, inches, etc. A table of frequently used values is provided in one of the appendices (see Table A–1). Note that the formula is more often used to compute the required character height, rather than the visual angle.

The second angle shown in Figure 2–1 (5°) yields the maximum size that will allow a viewer to grasp a piece of information in one eye gaze. This maximum is usually somewhat less of an issue but nevertheless can and should be applied when possible.

Another area deserving of our attention is the use of color. This is a notoriously difficult area in HMI design. Designers often rely on colors to convey information, but sometimes fail to realize their limitations. To start with, about 8% of men and 0.5% of women suffer some form and severity of color blindness. Therefore, even using "bold" colored text or coding information with color will not work for users unable to discriminate between colors. A few simple design rules will, however, allow the designer to get around these difficulties while preserving the important benefits provided by color coding:

- Do not write text in color; if you absolutely need to use color and text simultaneously, color the background of the text rather than the text itself.

- When you design a display, make sure that it will be usable in monochrome; once this is achieved, use color to enhance it. This is a controversial statement, but the controversy being somewhat philosophical in nature, we stick to this recommendation.

- If you use color as a means of coding visual information, then always use secondary coding. For example, use color and shape, or color and position, or color and border.

One last area of visual perception we will address is the effect of the viewer's age. Unfortunately, it is a fact of life that as people age, their visual acuity declines. For our present purposes, we will simply mention that older operators tend to require larger character height (within the 16 to 22 minutes range) and also tend to require more lighting than younger operators.

Memory

An extended discussion on memory is beyond the scope of this chapter, but we will briefly discuss the concept of working memory as it relates to HMI design.

To hold information required to carry out their immediate tasks, operators will use what is known as working memory. This short-term memory has a number of features, such as the ability to briefly retain information by "rehearsing," as is the case when one reads a phone number in a phone book, then "rehearses it" in one's mind while dialing the number.

Working memory has limited capacity, which is usually expressed as seven plus or minus two pieces of information (a piece of information, in this context). It is this limited capacity that must often be taken into account while designing an HMI. The issue, obviously, is to determine what is considered to be a piece of information. A short answer is that a piece of information is something that makes sense to the operator and stands on its own. For example, the word "plant" could be seen as a single piece of information, thus implying that between five and nine such words could be remembered by the operator for a short period of time. However, the serial number *75a49* could be considered as five chunks, as there is no global meaning to the string of characters. (The skeptical reader is invited to open a phone book for a foreign area, then try to remember more than one unfamiliar phone number).

The design implications of limited short-term memory are simple: do not force the operator to remember more than five to nine chunks of information; if needed, repeat the information elsewhere, wherever it is needed on the HMI. Also, since older operators tend to be more susceptible to disruptions to the content of their working memory, HMI designers should err on the side of fewer required chunks of information.

Decision-making

People perceive information, then remember what is needed to carry out their tasks, and then make decisions. This latter topic is very broad, and we will limit ourselves to some key elements. The Skill-Rule-Knowledge (SRK) model of human decision-making, developed by Jens Rasmussen (Rasmussen, 1983), a well-known researcher who did extensive work on operator modeling while he was at the Riso National Laboratory in Denmark, has become perhaps the most popular tool to understand and explain operator behavior in recent years. Briefly, Rasmussen's model divides human behavior as:

- **Skill-based behavior (SBB)**, which corresponds to the nearly automatic response of an operator handling well-known situations. A key thing to understand is that this type of behavior produces the best performance in terms of speed, accuracy, and error rate.

- **Rule-based behavior (RBB)**, where the operator follows a reasonably well-known process and/or executes procedures (written or not). Performance in RBB is usually very good, although not as fluid nor near-perfect as in SBB.

- **Knowledge-based behavior (KBB)**, where the operator must resort to his or her fundamental knowledge of the process, or of other areas, to solve a problem. KBB tends to be difficult and error-prone, often by an order of magnitude more than RBB. This is the riskiest and slowest means to carry out a task.

It should come as no surprise that when an individual engages in a novel task, he or she normally acts in KBB fashion. Over time, the individual will slowly migrate to RBB and, in some cases, to SBB. Therefore, we wish to train operators in the latter two types of behavior.

One last point implicit in SBB, but even more in RBB, is that most people—operators, engineers, technicians, doctors, plumbers, electricians, firefighters, cooks, etc.—tend to re-use, to the largest extent possible, recipes that have worked well in the past. Typically, decision-making will include identifying known successful recipes that correspond most closely to the present situation, adapting the best-matching recipe, simulating it mentally, and then implementing it. Only when no existing recipe exists will the individual create one through KBB, with its associated difficulties and potential errors.

Internal or Mental Model

The notion of *internal* or *mental* model refers to the way people understand elements of the world with which they have to deal. For example, imagine you have to explain to a child how a car can accelerate. The easy answer is that this can be demonstrated by depressing the accelerator. However, further questions by the child (a common tactic) may lead to attempting to explain that the accelerator changes the amount of fuel reaching the combustion chambers in the engine. Unless the person doing the explaining is well versed in engine mechanics, this is often an area where things start to get fuzzy. Is it the increased fuel flow that draws more air into the combustion chamber, or is it the air flow that draws the fuel? And then, what about combustion? If the explainer does not know that combustion is essentially an oxidation process, the explanation will become increasingly fuzzy, if not outright wrong.

This simple example shows we all have mental (internal) models of the world around us. More importantly, mental models have a number of characteristics that must be understood to avoid costly mistakes in HMI design. Some key characteristics are:

- Mental models used by an individual are typically his or her own and differ, in general, from those of another individual. These differences often concern some critical aspects of the model.

- The mental models of an individual are often incomplete, even for very experienced persons. In the car example, even seasoned drivers are often unaware of the many mechanical processes involved. One reason is that drivers rarely need to know about those processes.

- Mental models are often goal-oriented—that is, they are used to support an individual seeking to achieve a goal.

Therefore, it is almost always a mistake to rely on a single individual—even an experienced person—as the sole or main source of knowledge for specifying an HMI's content.

Human Error

Our last topic in understanding the user is one dear to HMI designers: reducing human error. As important as it is for designers, organizations, and regulators, this goal has proven elusive for a number of reasons. Human error has been credited as a contributing factor in most accidents that have occurred in high-risk domains. To try to understand why human error occurs, it is first necessary to identify the types of errors that can be made.

One popular error taxonomy, which has proven to be a useful tool to reduce human error, classifies errors as being either *slips* or *mistakes*. A slip (or error of omission) corresponds to the case where the operator carries out a well-known task but somehow omits one of the steps or performs a wrong one. This is similar to the car driver getting off at the wrong highway exit and ending up at work, rather than at the shopping center, on Saturday. Slips tend to be associated with SBB and RBB decision-making. Mistakes (or errors of commission) occur when either the operator does not know how to do something, and must therefore improvise, or when the environment—for example, the information provided by the HMI—leads him or her down the wrong path. It thus follows that mistakes are frequently associated with KBB.

The design implications of this understanding of human error are numerous:

- Slips may be mitigated by building error-prevention mechanisms, such as confirmation dialogs or the proper location of interactive devices, into the HMI.

- Mistakes may be mitigated by providing decision support functionality, ensuring that information provided by the HMI is reliable, and training the operator to distinguish between normal and abnormal situations.

Another tool for approaching human error in complex systems has been to use a notion known as latent pathogen (Reason, 1990). A latent pathogen is an error, be it technical, social, or otherwise, that exists in a dormant state in a system and is normally well-tolerated by the system. Examples of latent pathogens are deficient operating procedures, plant configuration errors, and risky, but usually tolerated, operating practices. Studies have shown that most grave accidents have involved several latent pathogens that have been "awakened" by a specific situation and have then created a causal link leading to an accident.

Latent pathogens originate from a number of sources, but have most often been related to maintenance activities in process control applications. There are various means used to control latent pathogens—such as work planning and control for maintenance and continuous improvement programs—and, where possible, to eliminate latent pathogens on a continuing basis. Another means, known as safety culture[1], has been introduced in the nuclear work domain and is being adopted in a number of other high-risk industries (e.g., medical, aeronautics). A detailed explanation of this concept is beyond the scope of this book but, suffice it to say, it actively encourages a proactive approach to continuous improvement. It fosters the active engagement of staff at all levels to enhance operations, in particular by identifying and removing technical and organizational problems even though they may not have yet led to an incident or accident. A good safety culture is a powerful tool in achieving a safer and more productive facility by actively searching out and removing latent and active problems, including organizational issues such as poor communication or work coordination.

---

1. Defined as "The assembly of characteristics and attitudes in organizations and individuals which establishes that, as an overriding priority, protection and safety issues receive the attention warranted by their significance." Accessed on 17 June 2005 at: http://www.ns.iaea.org/tutorials/regcontrol/intro/glossarys.htm

Operator Monitoring

Understanding how an operator monitors a process may seem like both a trivial and an almost-impossible task. It might seem trivial because operators know what they are doing and monitoring the process is just one of their many jobs. On the other hand, if one stops and looks at how they actually do it, then this seemingly simple task turns into an unwieldy assortment of smaller tasks from which it appears difficult, if possible at all, to extract any kind of a pattern. Yet, the implications of understanding how an operator actually monitors the process are profound. In fact, the alarm-handling capabilities of some classes of modern distributed control systems are a legacy of earlier studies that had concluded, perhaps somewhat prematurely, that operators wait for an exception to occur in the process and then act to rectify it (Zwaga & Hoonhout, 1994). More recent studies have identified that two types of behaviors are to be expected from operators: management-by-exception and management-by-awareness.

Management-by-exception refers to the behavior explained earlier, where operators wait for an exception to occur and then deal with it. This has the advantage of reducing to almost nil the effort expended to monitor the process. However, it includes the inconvenience of becoming out of touch with the process and not being able to avert a problem before it becomes quite noticeable.

Management-by-awareness, on the other hand, implies that the operator will continuously monitor key variables in an effort to get a feel for the process and ensure that it runs smoothly. The advantages of this approach are obvious. For example, it is easier to catch a developing situation before it becomes unmanageable, and the alarm floods typical of modern annunciation systems can more often be prevented. On the negative side, this approach requires a very dedicated operator and is somewhat difficult to practice when the process is unstable.

In most cases, it is expected that well-motivated and well-trained operators will use a mixture of these approaches and that management-by-awareness will tend to be more widely used when the process is more stable, while management-by-exception will be more appropriate during times of transition. Obviously, individual differences will continue to occur. Therefore, what good does it do for us to know about these behaviors? From an HMI design point of view, it is important to realize that both behaviors need to be supported. This leads us to identify two key HMI design principles:

- Information required to manage the process by exception should always be visible to the operator. Pragmatically, this is because the information is usually very simple (e.g., good/bad indicator, alarm message) and it requires minimal real estate to be displayed.

- Information required to manage the process by awareness should always be available upon request by the operator. Contrary to the information required to support management-by-exception, the information required here tends to be associated with trend graphs and will occupy more real estate. It should thus be possible to call it up whenever the operator wants to manage the process by awareness.

Linking HMI Design and System Design

Having improved our understanding of users, we will now try to understand better how HMI design fits into the overall system design cycle, as shown in Figure 2-2.

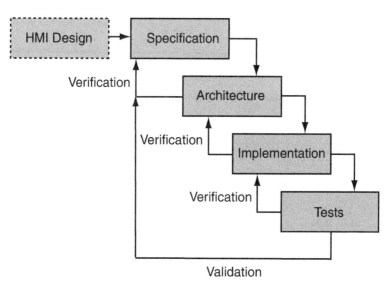

Figure 2-2 *HMI Design and the Overall Design Cycle*

Figure 2–2 shows a typical waterfall process, from specification to architecture to implementation to tests. It might be puzzling that HMI design is shown ahead of the overall system design process, yet it is crucial to understand that this is the optimal time for it to occur. It has been argued and demonstrated that most of the errors in real-life projects, and the costs to correct them, are associated with the specification of what needs to be designed (Cooling, 1991). Any improvement in the quality of what needs to be designed, implemented, and tested will thus provide immediate benefits that will multiply with each step of the overall system development cycle. Another way to look at it is to realize that virtually all design projects that include either a significant number of interactive components, or that introduce new display or interactive functionality, tend to be associated with a large portion of uncertainty, and that HMI design aims at reducing this uncertainty.

Having a better understanding of where HMI design stands, we can now look at how we will carry it out. Several situations can occur, so we will start with the one that entails the most complex issues—that is, starting from scratch for an arbitrary system.

Standards and Guidelines

Several standards on how to design HMIs have been developed in various domains over the years, and we will rely heavily on those. Table 2–1 lists standards, guidelines and recommended practices that are of particular importance in HMI design for process control applications.

The standards, guidelines, and recommended practices in Table 2–1 represent the collective wisdom of a large number of experts in various specialties of human factors engineering. However, as useful as they are, they are grossly insufficient to design good quality HMIs. The reasons for this are numerous, but can be summed up simply: most standards address the issues of the attributes of individual components (e.g., how large should the font be, what is the best color for a particular indicator); they do not, and generally cannot, address the central question of what are the interactive functionalities the operator needs to achieve the purpose of the operation for which he or she is responsible. The answer to this question requires using process-oriented standards, that is, standards that describe the process of designing a suitable HMI. This is the subject of the next section.

Table 2-1 Standards and Guidelines Relevant to HMI Design

Document	Comments
Human-System Interface Design Review Guidelines—NUREG-0700, Rev. 2	Very extensive set of guidelines from the nuclear industry, but with wide applicability to industrial process control. Covers HMI components, workstation design and control room design.
ISO 9241 Ergonomic requirements for office work with visual display terminals	This 17-part standard was originally intended for office computerized workstations, but the widespread use of commercial-grade software and hardware in industrial applications has turned it into a "must know" source of information. The various parts of this standard cover information presentation, menu design, form-filling, keyboards, displays, and the like.
Mil-Std 1472F Design Criteria—Human Engineering	Another extensive source of knowledge, guidelines and principles of HMI components and workstation design. Very thorough; covers both conventional and hardwired panels, switches, etc., as well as computerized devices.
ISA-5.5-1985 Graphic Symbols for Process Displays	This ISA standard provides a list of industry-accepted graphic symbols for process control applications.
EEMUA 191 Alarm Systems—A Guide to Design, Management and Procurement	This guide from the Engineering Equipment & Materials Users' Association (a European-based, non-profit distributing industry association [www.eemua.org]) is considered to be the leading source of good practices with respect to alarm management techniques.
http://www.hse.gov.uk/humanfactors/comah/alarmhandling.htm	This website provides an abundance of information and guidance on best practices for alarm handling.

Table 2-1 Standards and Guidelines Relevant to HMI Design (cont'd)

Document	Comments
Official Guidelines for User Interface Developers and Designers (for XP and previous versions of the Windows operating system) *Windows Vista User Experience Guidelines* (for the Vista operating system) *Java Look and Feel Design Guidelines, Second Edition*	These, and equivalent documents for other popular computing platforms, provide invaluable information on the most detailed features of individual components of computer-based HMIs. They provide extensive guidance on the use of individual widgets, windows management and the like. They are usually available for download from the Web sites of the organizations that provide the software platform.
ISO 13407:1999 Human-Centered Design Processes for Interactive Systems	This standard is widely applicable to the design and specification of interactive systems. Even if somewhat vague as to the specifics, it provides the soundest basis for the process used in designing HMIs.

Overall HMI Design Process

For reasons of practicality, and to have a sound foundation, both theoretically and pragmatically, the method we will use will be based on ISO 13407. Figure 2-3 describes the main phases of the process mandated by this standard.

We briefly describe the overall process here to provide a general overview. A more comprehensive description will be provided, using an example we will describe shortly.

The first phase is a planning stage, during which a number of decisions are made. For example, how will the outputs from the process feed into the overall system development cycle? What will be the schedule? Who will be involved, and how will they participate?

The next phase is to specify the context of use, that is, to determine who will be the users of the future system, what their characteristics are, what tasks they will be expected to carry out, and what the relevant elements are of the environment they have to deal with. This is usually not as trivial as it may sound, and it may even sound fuzzy. This is actually not the case. Several techniques can be brought to bear on this phase.

20 Chapter 2—Designing a New HMI

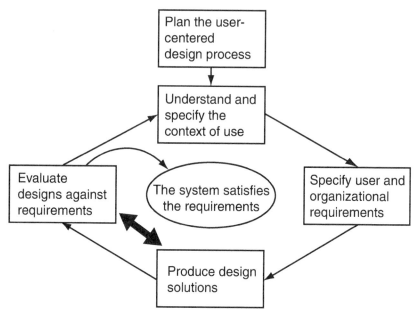

Figure 2-3 *Overall HMI Design Process*

Specifying user and organizational requirements is next. This entails identifying the required operational (and sometimes financial) objectives of the organization, the needs associated with communication and cooperation among the users, and the relevant aspects of work organization. This is also a good time to identify training and support, as well as maintenance assumptions.

The next phase is to produce design solutions. This is an interesting, yet dangerous, part of the process, as this is a place where things can easily steer out of control. ISO 13407 describes a process that is sometimes referred to as being *user-centered*. People may take it to mean that the user will produce the design, perhaps with some help from the designer. This is a mistake, much in the same way as it is to ask an experienced car driver to design a new vehicle. Chances are that the resulting vehicle would be fun to drive for a short while, until it encountered conditions unanticipated by the would-be designer; an oversized engine or brakes (depending on the driver's inclination) and a lack of standardization could also be expected. In short, we argue that designers should design and that operators should operate (designers also make poor operators).

Producing design solutions will rely on a variety of methods and approaches. We will explain in the example to follow that an architecture-based approach usually yields satisfactory results. Also, even though designers will try to create the best design based on available information, it must be realized the design process will be iterative and corresponds, in fact, to a gradual reduction in uncertainty with respect to the content of the HMI and its organization.

Evaluating the results against requirements is the next logical step of the HMI design cycle. This can be carried out through a variety of means; however, usability testing is normally the method of choice. It consists of carrying out simulations of selected tasks using representative users. Measures of performance are taken, and the design is refined based on the results from the usability tests. Figure 2–3 shows the results from the evaluation as feeding either into the analytical phases (if the requirements are not met) or into the system (or software) requirements document. This must be considered carefully. The relationship between the production of design solutions and the evaluation of those solutions against requirements is very close. Typically, a mock-up will be subjected to usability tests for a few hours and the lessons learned will be incorporated into a revised version of the mock-up (rather than being used to re-visit the analysis phases) in a matter of days, and sometimes even hours. Doing this appears to bypass the analysis phases, which we should normally re-visit if the design fails to meet the requirements. Reality is somewhat more subtle. Normally, the results from the evaluation will be carefully considered and elements of the analysis phases will be revisited only if there appears to be a problem in the analysis that will prevent successfully arriving at an enhanced mock-up. Otherwise, we will pass through the analysis phases and will incorporate the lessons learned into the new mock-up.

A few points of interest for this process:

- There are no hard-and-fast rules as to how many design-evaluate iterations will be required, but for a moderately complex system, or a sub-set of a very complex system, one should start to converge toward a stable design after three to five iterations.

- Failure to converge after a few iterations often points to underlying issues in one of the analysis phases. The author's experience is that when the design fails to converge, there is often a problem in the definition of the mission of the system. It is often too vague or does not represent the objectives of the stakeholders.

- Once the usability tests indicate that the requirements have been met, the results—usually in the form of a comprehensive mock-up, and often accompanied by supporting documentation—are fed into the system or software engineering requirement specification.

Much more could be said about the process. Rather than going into a more elaborate discussion here, we will look at a simple example that will serve as a vehicle to explain the process. The example will also be used to introduce the various techniques and tools that are useful to implement the process.

Detailed Description of the HMI Design Method

In this section, we will look at details in the various steps of the HMI design method that we introduced earlier. As we will see, the method can be used irrespective of the underlying technology (SCADA, DCS, hardwired panels, etc.) To enhance understanding, we will simulate the implementation of each of the steps in the design of an HMI for a simple process monitoring and control system. We will start by describing the application that will be used for the example.

Example: Newsprint Papermaking

The example described here is based on a simplified and idealized version of a common industrial process, that is, a paper machine producing newsprint. There are commercial systems available that provide monitoring and control functionality for this type of process. For the purpose of the discussion, however, we will assume that those responsible for implementing computerized process control have decided to design their own custom HMI. Figure 2–4 shows our fictitious newsprint machine.

The main components of the process are:

- The headbox, where the stock (i.e., the slurry made of up to 98% water and about 2% wood pulp) is stored and then distributed as evenly as possible on the wire.

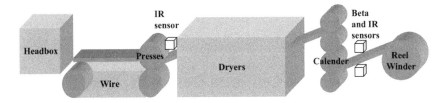

Figure 2-4 *Fictitious Newsprint Machine*

- The wire transports the stock towards the presses; meanwhile, vacuum pumps start to extract water from the stock.

- The paper sheet is approximately 60% water and 40% pulp after it has passed the presses. It is then sent to the dryers (rotating, steam-filled cylinders).

- Exiting the dryers, the sheet is ready to be calendered (to improve its hardness and finish profiles) and has a moisture content ranging typically between 7% and 10%.

- Once it has been calendered, the sheet is wound into a cylinder at the reel winder, from which it will be sent further down the production line to be cut and packaged.

This is admittedly a very rough and incomplete description of the process. However, it will suffice for our purpose.

There are a number of analog and digital measurements taken to monitor the process. For example, beta and infrared (IR) sensors measure some of the final parameters of the process while conventional transmitters provide levels, pressures, etc. Control actions are applied to a variety of devices, such as the stock valve that ultimately controls the weight per area of the sheet (e.g., 30 pounds per ream), and the steam valve that controls the flow of steam through the dryers, thus changing the amount of moisture remaining in the sheet at the calender.

There are quality, productivity, and safety goals that have to be achieved for this process as for any other process; we will detail those as we go and as it becomes necessary during the course of designing the HMI. With this general understanding of the application, we will design and test an HMI that is suitable for optimizing the achievement of those goals.

Planning the User-Centered Design Process

First, we describe the planning phase, then we give an example of how it can be applied to our fictitious application.

Description of the Planning Phase

As discussed in the overview, planning the user-centered design process is the first step, and entails looking at a number of issues to help ensure that we get the desired benefits from the effort that will be expended. For our example—which also applies to virtually any application—we need to:

- Determine the timetable governing the delivery of the specification of the HMI design.

- Identify subject matter experts who will be consulted and ensure that they will be made available. Their involvement will normally be limited to an "as requested" basis.

- Identify which documents will be produced, agree on the review and approval process that will be used, and agree how the content of those documents will feed the system development process.

- Determine how many operators will participate, and how.

A common mistake is to gather too large a team to participate in the design of the HMI, sometimes in an attempt to reach a consensus on what the content of the HMI should be. This is generally not necessary, as a small team will achieve the results quicker and will be more cost-effective, while achieving project objectives if they use the design process recommended here. Typically, the core HMI design team will be one or two trained people who will only need the support of other staff as they progress through the analysis, design, evaluation and specification phases.

There are a number of other ingredients that can, and often should, be included in this phase that will contribute to a better understanding of what needs to be done and will help ensure that the process is successful. For instance, it is often beneficial to organize short and somewhat informal training sessions to explain the HMI design process to management and operating staff. Those sessions are especially well-received if it is possible to identify an existing example of an HMI that has been troublesome to the operating staff, such as a poorly designed panel or computer display, and to explain how the HMI design process will help prevent those problems. A nice, and often unexpected, side effect of these meetings is that requests may then be made to apply the HMI design process elsewhere in the plant.

Another ingredient worth mentioning is that the use of a mock-up, which will prove invaluable as a means of communication with the various stakeholders.

Example: Carrying Out the Planning Phase

In this example, we have been asked to provide the functional requirements for the user interface for a new piece of software that will be either procured and configured, or built in-house. This new piece of software will have an HMI that has to support the operator in monitoring and controlling the process in both normal and abnormal, but expected situations. There is a catch, which is that our portion of the requirements is to be produced within a month from the start of the project. Furthermore, since this is a novel way to do things, project management will heavily rely on our input to determine what needs to be done and when.

There are a number of things that need getting done. We have just been informed that our output has to be made available to the engineering group in one month, so we need to work the schedule backward. We decide to solve the problems one by one:

- We know about HMI design, but we do not know much about the domain nor the process. We will need access to a subject matter expert. Ideally, we would like to have access to a trainer, but none is available. We will have to rely on a newly promoted foreman. Given the timetable we have, we expect to ask for a few hours of his time, probably something in the vicinity of eight to ten hours.

▪ Since we will feed our results to the engineering group, we will need to set up an appointment to discuss how they will use those results. There is also a need to agree on the level of formality of the requirements we will provide to them (i.e., to define their expectations in terms of formality and content). Given the good relationship that we enjoy with this group and looking back to a previous project, we will suggest to them that we will provide a short design description document that they can use concurrently with the latest mock-up that will have been tested. This approach tends to work well, and it minimizes the paperwork that has to be generated, yet it preserves a certain amount of formality.

▪ The next order of business is to secure the availability of some of the operators to participate in the evaluation of the HMI mock-ups. Since the project is simple and since we have some knowledge of comparable systems, we expect to be able to complete a good HMI design with three rounds of design-evaluate iterations; however, we will plan for four rounds, in case something unexpected happens. With an average of five operators per round, and an expected one to two hours per operator, we will need to set aside about thirty hours of operator time for evaluating the HMI design and about ten hours for obtaining inputs for the analysis activities.

With the above information, we can draft a simple plan for the HMI design project. Table 2–2 shows the steps that have to be taken as well as the effort that we are planning for.

The schedule is definitely aggressive, but not unrealistic. We even have one last "spare" iteration for design-evaluate.

Understanding and Specifying the Context of Use

Here again, we will explain this phase and then demonstrate through an example how it can actually be carried out.

Table 2–2 HMI Design Project Schedule

Step	Week 1	Week 2	Week 3	Week 4
Specify context of use	■			
Specify user and organizational requirements	■			
Produce design solution		■	■	■
Evaluate design solution		■	■	■
Provide specification of the final solution				■

Description of the Context of Use

This first step in the analysis is fairly straightforward. During this phase of the work, we aim at:

- Identifying the mission of the HMI. This is the most important step of the whole process and will have a profound impact on the cost and performance of the future HMI. The mission should be a single statement, so clear that different individuals will have the same interpretation. The statement, "The new HMI will provide a graphical interface to the operator for managing the process" is next to useless. Providing a graphical interface does not lead, by itself, to any enhancement in the management of the process. However, a statement such as, "The new suite of computerized displays for the digital control system will provide the monitoring and control capabilities required by the operators to ensure that the process remains within the productivity, quality, and safety bounds defined for this process, both for normal and for the following abnormal situations (list of situations)" is unambiguous. It tells us exactly what is required from the design.

- Identifying who the users of the interface will be. For example, will it be used by operators only, or will the plant manager need some displays as well? What about the instrumentation and control (I&C) technicians? Will the process engineer need some support for tuning the control strategies? Once we know the types of users, we need to determine the characteristics of those users that will have to be considered in designing the new HMI. This could be a long and tedious process, so we will attempt to be pragmatic. Using a predefined list of inquiries, as in Table 2-3, will often yield the necessary information.

- Identifying tasks that have to be supported for each of the types of users. This is both intuitive and unfamiliar. A key thing to understand is that the HMI is there to help the user carry out certain tasks; therefore, identifying those tasks is the number one issue to resolve in order to specify the content of the displays. Identifying those tasks implies more than creating a simple list of tasks. It may come as something of a surprise that users rarely, if ever, do what we think they do. Listing their tasks means observing them as they carry out specific tasks. Designers are always amazed at the ingenuity operators deploy to start up a system, change operating settings, or deal with alarms. The better we understand the user's tasks, the more productive the new HMI-user combination will be. The process of understanding the user's task is known as task analysis; one of the best techniques to use to carry out a task analysis is *Hierarchical Task Analysis* (HTA). The steps are:

    (1) Identify a list of tasks that must be supported by the HMI. Typically, this will include tasks required to start up, adjust operating parameters, shut down, deal with expected abnormal situations, etc. These tasks can often be identified through a discussion with operation management or with senior operators.

    (2) Select a subset of those tasks for which the information or control requirements are not clearly known, or which are critical to the safe and efficient operation of the plant. Those tasks will be the subject of the actual HTA.

    (3) For each of the tasks in this subset, observe operators carrying out those tasks. Be careful to make several observations with operators of various proficiencies. Be

Table 2-3 Examples of User Characteristics

User characteristic	Bearing on the design
Familiarity with the tasks.	The higher the familiarity with the tasks, the lower the need to incorporate task-oriented support into the HMI.
Familiarity with the type of user interface devices that will be used.	A surprisingly high proportion of users have very limited mousing skills, and keyboard proficiency is often a rarity. On the other hand, if users are proficient at using modern software packages (e.g., word processing, electronic spreadsheet), one can often leverage this knowledge and shorten the learning curve for using the new HMI.
Handicaps to accommodate.	Do we need to take into account that some of the prospective users will have mobility challenges, or will have other functional limitations? This will need forethought to ensure that the new HMI will support those users.

especially wary of the tendency to ask one senior person to tell you how this task is normally done. This invariably leads to incomplete comprehension of the task.

- Identifying software and hardware devices that the user of the new HMI will have to use or interact with. For example, if the new HMI will use commercial-grade hardware and software, this needs to be specified, as it will heavily influence the design options available to the designer.

- Identifying standards and guidelines that have to be used or followed. Some of those standards may be linked to legal requirements while others aim at ensuring consistency of operation.

- Identifying other constraints that have to be taken into account, be they technical or even legal.

Example: Analyzing the Context of Use

Moving forward with the example we have used so far, we decide that our first technical objective is to define and ensure buy-in of the mission assigned to the HMI. We believe a fair statement of that mission is: "To achieve a suitable level of usability to effectively support operator monitoring and controlling of the process during commissioning, routine start up, normal and expected abnormal operation, grade change, and shutdown." We will further refine the meaning of "suitable level of usability" as we progress, but the current mission statement determines exactly what we will do.

Our mission definition takes us part of the way in identifying the types of users the new HMI will support. Discussing it with project management, we realize there is an expectation to support:

- The main operators for the process; this is already covered.
- The instrumentation and control technicians, for tasks such as calibration and troubleshooting.
- The process engineer who, from time to time, will have to tune the regulation algorithms on the system.
- Last, the ubiquitous boss is not to be left out, so some aspects of the HMI will need to be tweaked to accommodate his or her needs.

We recognize that, in some cases, technicians, process engineers, etc. may use HMIs dedicated specifically to their needs, with the operator HMI designed purely for the operator.

However, the most important user group will be the operators, since they basically control production. Other users will still be important but, even though we will strive to support them as best as we can, they will clearly have a secondary role. Given this, we decide to add the following to our mission statement:

"To achieve a suitable level of usability to support, in order of priority:

- The operator and other operating staff in monitoring and controlling the process during initial commissioning, routine start up, normal and expected abnormal operation, grade change, and shutdown;

- The instrumentation and control staff in diagnosing problems and ensuring that the calibration of the system is satisfactory; and

- The process engineers in ensuring that regulation algorithms perform adequately."

We are pretty happy with this definition and decide it will do for the moment. We then proceed to produce a *task analysis* for the users. While the technique is said to be simple, many have tried it and given up. Fortunately, a colleague who has successfully used this technique for an operators' training course has offered some advice.

Following his suggestion, we start by identifying a series of tasks the operators must carry out to achieve the mission identified previously. To do this, we decide to focus on normal operations and ask a senior operator to describe in broad terms the main steps an operator carries out to start up, operate, and shut down the paper machine. He tells us a series of steps that are typically taken. To ensure that we have captured the right information, and all of it, we make it a point to go and observe the operator carrying out those tasks in the plant; we identify a few discrepancies with the previous description of typical steps and we discuss those with the operator. He explains that those are special cases that sometimes occur and agrees that they are part of what an operator is expected to know. We then repeat the same process with another senior operator and obtain another series of steps; again, we carry out some detailed observations in the field and identify minor discrepancies that we resolve through a discussion with the second operator. It is up to us to represent those steps in terms of tasks. Our colleague has suggested expressing each task as a verb and a noun (e.g., "increase speed"). Also, remembering that a task has a start point and duration (or an end point) helps us to stay on course.

A task can generally be broken down into sub-tasks. This introduces one troublesome area when an operator says such things as, "When this or that happens, you have to do that, but if something else happens, you do this." This leaves us a bit puzzled as how to best capture this knowledge, but our colleague then judiciously explains plans. In short, a plan describes how sub-tasks that make up a task are grouped together; in his words, "Think of a plan as the recipe (the operator) follows to carry out a task." Figure 2–5 shows a fragment of a model of the operator's task.

We now review our task diagram after organizing the material we collected and we see that:

- The top task happens to be an excerpt of the mission for the new HMI.

- The main tasks, supporting the top task, were identified through our discussions with senior operators and have been organized to give a global picture of the situation. The resulting diagram is hierarchical and easy to read. Upon being told by his or her supervisor, the operator starts up the paper machine (task 1). This task can be broken down into sub-tasks (those sub-tasks are not shown here). Once the machine has been started up, the operator will monitor it to ensure that the operational goals (safety, production and quality) are being achieved (task 2, and sub-tasks 2.1, 2.2, and 2.3). Simultaneously, the operator will make whatever minor adjustments (e.g., speed, sheet basis weight) are needed to ensure near-optimal operation (task 3). Should monitoring reveal that a goal is being threatened, the operator will take the required remedial actions (task 4). If a grade change is required, such as going to a heavier paper, the operator will carry that out (task 5). Eventually, perhaps to carry out planned maintenance activities, the operator will, upon instruction from shift management, shut down the paper machine.

- We had to go back a couple of times to each operator after each correction to our diagram, but we now feel confident the resulting model is an accurate representation of what operators do, and, therefore, of their information and control needs.

- As a final step, we take advantages of current operations in the mill and observe a few start-ups and shutdowns, as well as current operations. We do a few minor touch-ups to our task diagram and now feel confident it is sufficiently sound to form the basis of defining display contents and navigating through the HMI.

Understanding and Specifying the Context of Use 33

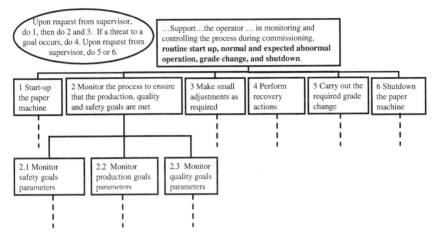

Figure 2–5 *Fragment of a Model of the Operator's Task*

- It should be noted that a more thorough treatment of HTA is beyond the scope of this book, but very good descriptions are available elsewhere (e.g., the one found in Kirwan, B. & Ainsworth, 1992).

While we were drafting our task diagram, we had sent some requests to determine which software and hardware platforms would be used for the HMI. The reply was that commercial-grade PCs—each with a 21-inch monitor and a standard keyboard and mouse—would be used, mostly to simplify the long-term support of the workstations. This should be sufficient for the job, but we make a note to ensure there will be sufficient room on the console supporting the mouse to support right- or left-handed operation. The software platform will be a popular, commercial-grade operating system with a graphical user interface that is also used for running the office automation software at the plant. This rings a bell for us, though, as we realize it would be nice to capitalize on whatever computer operation skills the operators may have acquired with software packages currently used in the mill. This, in turn, leads us to specify that the new HMI will have to conform to the style guide for the platform; style guides are usually available from the provider of the software operating system (e.g., Windows, Linux); sometimes, style guides also exist for some languages (e.g., there is a style guide for Java). The style guide dictates the HMI's look and feel and, therefore, has an important role to play.

We also noted, while talking to project management, that the monitors and other components making up the HMI will be housed in an air-conditioned room where the operator will perform a fair portion of his or her work. Again, we have no particular reason to question this choice.

At this stage of the game, and given that this is a new endeavor for us, we believe we have determined as much as we can for this phase of the project.

Specifying User and Organizational Requirements

In this phase, we aim at completing our understanding of the issues and constraints that have to be dealt with to design a high-performance HMI. As previously, we will describe how to analyze user and organizational requirements, and we will illustrate how this can be carried out through an example.

Description of User and Organizational Requirements

As mentioned before, analyzing the user and organizational requirements leads us to specify what the operational and financial objectives for the project are. We can often use those to identify criteria we will use to determine whether the new HMI, and even the new system, is successful. Examples of such criteria are:

- Decrease the number of operator errors on task XYZ.

- Decrease the time to start up the system.

- Shorten the training required to operate the system.

- Improve process yield by reducing quality-related rejects.

Each of those criteria should be meaningful and measurable. Otherwise, testing to determine whether or not we met them would be impossible. There should now be a reasonable number of these criteria. As a rule of thumb, three to five criteria is a good number. Remember, we will have to prove the design has met those criteria to demonstrate that it is satisfactory. It should now be obvious that proper HMI design is more concerned with the performance of the human-machine system than it is with the users' subjective preference for one design over another.

In this phase, we also need to clarify or specify the assumptions regarding the cooperation and communication needs of the users. Will the workstation be used by more than one operator? Is there a need to provide displays that will support coordination between two users carrying out a given task? Are there communication requirements that have to be met so the user will be able to carry out his or her task(s) efficiently with the HMI?

A sometimes misunderstood step in this phase is to specify, or identify, what training will be assumed for the users. The reason for considering training at this time is that it interacts with the design of the HMI. Planning to train the user more extensively will lessen some of the required functionality in the HMI, but may be more expensive in the long run; it may also mean the system will be more susceptible to human error. It is thus necessary to define at this early stage the type of content that will be required for user training, without actually having to develop it just yet. For example, if the training will assume that the operator will be familiar with how information and controls are organized in the HMI after 30 minutes, the HMI designer had better take note of that. It is worth noting that the link between HMI design and training is actually stronger than that. For example, the task analysis carried out during the earlier phase will be instrumental in determining the content of the training program.

Another topic that needs to be addressed at this phase is that of maintenance requirements. It is important for HMI designers to determine which needs and means have to be postulated for maintenance. Examples of questions whose answers may influence HMI design are:

- Will there be redundancy in hardware and software planned for the HMI? If so, how will HMI functionalities use these assets in case of failure?

- How will the user detect a hardware or software failure, and how will he or she transfer to the redundant components?

- Will the new HMI provide support for the maintenance staff, and if so, what kind of support will it offer?

- How will we ensure that maintenance activities on the new system will not prevent the user from monitoring and controlling the process?

Maintenance requirements may have a profound impact on the actual design of various components. For example, if the new design relies on hardwired panels, good practice would dictate that maintenance-related access to these panels should be from the back to avoid interfering with the tasks of the operator. For another example, if the maintenance staff is to rely on support from the HMI, then this support will have to be designed in. A comprehensive appraisal of maintenance-related requirements must be conducted.

Example: Analyzing User and Organizational Requirements

We have a reasonable understanding of the context into which the new HMI will have to fit. We now need to understand which requirements it will have to meet from the user and organizational point of view. To do this, we have elected to have an interview with management about what objectives the new HMI is expected to achieve. We are a bit nonplussed to learn they have considered the HMI to be just part of a monitoring system upgrade and have never really realized that its purpose is to contribute to enhanced productivity and safety; we are only too happy to bring this point to their attention. We are quick to point out the benefits that are normally derived from a sound HMI design and suggest a few objectives to make them tangible for the project:

- *Training time.* Currently, it is expected that operators will have to undergo a substantial amount of training to learn to use the HMI to run the process. We suggest that a sound HMI design should be able to shorten the training time; without any previous experience to guide us, we conservatively suggest that a 20% reduction in the planned operator training time should be achievable.

- *Productivity.* At the moment, start-ups and sheet breaks[2] are two major challenges. Start-ups are especially difficult, as the

---

2. On a paper machine, a sheet break can occur for a number of conditions (e.g., speed mismatch between the various rotating sections of the paper machine, wide variation in sheet density (or *basis weight*), operator error) which may lead to a sub-optimal tensile strength, making the sheet easier to break.

tensile strength of the sheet is a direct function of its density. However, no process information regarding sheet density is available until the sheet has been measured by the beta sensor near the reel winder. In other words, information regarding sheet tensile strength has to wait for the sheet to be strong enough to be calendered (admittedly a challenging situation). Reviewing the logs from previous start-ups and betting on a few ideas about providing a predictive value of sheet density derived from available process measurements, and also expecting to reduce some human errors, we aim to reduce the average start-up time by 10%.

- ***Product quality.*** Production trials are carried out routinely on the process, and maintaining key quality parameters within the required bounds requires, at times, all of the combined abilities of the automation system and the operator. Several subtle decisions have to be made, based on guesses rather than facts. Furthermore, some of the guesses require the operator to perform calculations quickly. Reviewing quality issues and, based on our understanding of the operator's tasks based on our HTA work, we are quite confident that we can reduce quality-related rejects by 10%.

- ***Waste.*** Our observations of operators, as well as discussion with shift management, suggest that some waste could, and should, be eliminated. Once produced, paper is wound around a reel to such a diameter that it can be rewound and split into smaller rolls to fill customer orders. Since it is very annoying to run short when cutting down a large reel into smaller rolls, it was observed the operators tended to make bigger reels. Any excess paper is then sent to the broke pulper, or dry end pulper for recycling[3]. A small safety margin might be acceptable, but currently, it is believed the current margin is too high, suggesting that a sizeable proportion of the production is simply being recycled. We expect it to be reasonably easy to incorporate some simple decision-support functionality into the HMI that will lead to a reduction of, say, 50% of this waste.

---

3. The broke pulper or "beater" is often like a pit where scraps of excess or unusable paper are sent to be shredded and remixed with water while waiting to be reprocessed.

We are pleased we have been able to identify areas where the new HMI design will contribute to the effective payback of the system. Further, we are somewhat surprised at the additional credibility that has been gained for the HMI design effort, and we find it makes the design team more sympathetic to the work plan we have produced so far.

Documentation

In previous sections, we have pretty much completed collecting and organizing the information that will form the basis of the design phase. In the author's experience, it has proven valuable to document that information and have it signed off. Although not strictly required, this sign-off step has a number of advantages. It consolidates in a central place all of the knowledge (or at least the documented knowledge) that has been accumulated so far. By circulating this document, its content and assumptions can also be rendered explicit, and various stakeholders will have a chance to agree to, or challenge, the information collected. At the risk of sounding pedantic, if the stakeholders do not agree with the results from the analysis phase, what good is it to carry out the design phase?

Once signed off, this document (the author often refers to it as the analysis document) should be controlled in such a way that inadvertent changes to it are not made. It becomes in effect a contract between the HMI designer and the stakeholders.

Producing Design Solutions

This section describes how results from the analysis work done in earlier phases crystallize to lead to a suitable HMI design. As mentioned earlier, the design phase is the fun part of the whole exercise, but also the phase where the designer or design team must exercise the most restraint.

While previously we restricted ourselves to collecting and trying to understand information, in this phase we will attempt to create a solution to a problem. This is quite a difficult process, and it warrants a few explanations. To start with, the whole notion of how we will design this new solution is somewhat fuzzy. Traditionally, designing an HMI for process control applications has relied on a mixture of the following:

- *Using the set of piping and instrumentation diagrams (P&IDs) corresponding to the system of interest as the content or representation for most of the displays or panels.* This is possibly the easiest approach to "designing" the HMI. The advantage is that the plant staff may be familiar with the P&IDs. The disadvantages, however, are numerous. First, it must be realized that the P&IDs were initially designed for supporting instrumentation and control needs and normally contain other details not related to normal operation. Thus, they often provide too much I&C information to operators, while being relatively poor in the information required to support the operation. For example, the P&IDs will rarely provide information regarding the achievement, or the causes for the lack of achievement, of operational goals. HMI components based on P&IDs also tend to be dense, so task-oriented information may be (and often is) hard to find as soon as the number of HMI displays becomes even moderately large.

- *Mimicking the HMI of an existing application.* This practice has the advantage of using something known to have worked elsewhere. The drawback, however, is such a design is very rarely, if ever, optimal. Since little systematic continuous improvement of HMIs is carried out after they have been deployed in real life applications, chances are this approach will fall short of expectations.

- *Purchasing a solution from a system supplier, to include a default set of HMIs.* Unfortunately, suppliers have very rarely carried out a systematic and well-grounded HMI design. As with the previous approaches, this strategy is also bound to lead, in most cases, to sub-optimal design.

- *In some extreme cases, the designer essentially relinquishes design responsibility by having senior operating staff indicate the content of individual displays.* The problems with this approach are simply too numerous to enumerate here, but suffice it to say that no user, no matter how experienced, has a complete and correct mental model of how a complex process operates. Further, users are good at using, but usually lack the training required to design.

Since we argue that the above approaches are not sufficient, then, surely, there must be a better way...and there is. The next sections will explain how to design an HMI that will, as much as possible, provide optimal or near-optimal support to the users.

Production of Design Solutions

Once we understand the context of use, and we have a workable set of user and organizational requirements, we have the basic ingredients to formulate a first mock-up of the intended system. Note the use of the word *mock-up*; at this stage we do not aim to produce any functioning code (in the case of a computerized HMI) or panel (in the case of hardwired panels). Rather, the mock-ups will be limited to what is absolutely required in terms of appearance to let the users go through the motions of carrying out simulated tasks (this will be discussed in detail in the next chapter). Actually, the first mock-ups are often best produced and tested on paper (this is true for both computer-based and hardwired HMIs). This has several benefits: it is very cheap to produce mock-ups and it is very easy and quick to change mock-ups, based on what we have learned. Mock-ups can normally be produced in a matter of a few days for the first ones, and modified in a few hours for subsequent versions.

One may wonder why we need to plan to modify the mock-ups at all, given that we have a good understanding of who the users are, what the tasks will be, which standards we will be using, and so forth. The unfortunate reality is that, even if our design is sound, little more than 50% of our first design can be expected to actually survive its harsh contact with the evaluation phase. This may sound like a discouraging perspective, but think of what the situation would be if the opposite held true: If we developed code for our first design solution, a significant chunk of it would be wrong. For the reader experienced in software development, this should ring a bell. The next question is: If we only get close to 50% of the final design, why bother with the analysis at all? The easy answer is: If we do not complete a good analysis, we will not produce an HMI that will even be close to 50% of a satisfactory solution.

So, how do we proceed with the design? There are two main phases that need to be carried out. The first is to sketch the *architecture* of the HMI. This will provide the global framework that will guide the rest of the design effort. This notion of architecture is familiar to engineers and technical staff; it simply means that, before designing individual components, one should have a precise overview of what the HMI will look like. Typically, an HMI architecture will include:

- A set of presentation conventions such as: How will we visually or audibly code normal and abnormal conditions? Which symbols will be used to depict systems and equipment? What will be the background color, etc.?

- A set of basic display types and associated layout templates to help the user to recognize where information is on a given display and to provide hints as to what type of display he or she is currently using.

- A specification of which types of help, or user assistance, will be provided, and how this will be implemented (electronic or paper format).

- A set of generic mechanisms for modifying setpoints and limits, for entering text (if required) and for retrieving information.

- An overall navigation map describing how the user will move about in the set of HMI information and control groups.

- A specification of physical resources that will be used. This specification will normally be based, in part, on information that is gathered when analyzing the context of use. However, some additional questions may surface, as we will see shortly.

Defining this architecture will also force HMI designers to think about the number of displays they expect to have to design, as well as how many monitors they will actually use. There are several principles that can guide how one goes about defining and validating HMI architecture. One of the first principles is that the main operator's tasks in process control have been shown to belong to a small set of generic tasks. Those tasks are:

- Monitoring the process to ensure that each of the operational goals (i.e., those related to productivity and safety) are, or continue to be, achieved;

- Controlling the process, thus enabling start-up, shutdown, changes of production throughout, or reconfiguring the process so as to produce a different type of product;

- Managing a process or system disruption;

- Diagnosing the cause of a threat to one of the operating goals; and

- Optimizing the process once the system is in a stable operating region.

Comparing these tasks with those shown in Figure 2-5 is a simple exercise because most process control tasks can be associated with one of a few generic tasks (Fiset, 2001). Those tasks have characteristics that bear directly on the design of the HMI's architecture. Key characteristics are provided in Table 2–4.

As stated earlier, most users' tasks in process control can be mapped into those generic tasks. There are a few ancillary tasks that cannot—for example, creating journal or log book entries or providing support to the I&C staff, such as during control loop tuning or calibration. They can be added on an ad hoc basis to the set of tasks that we need to accommodate.

Additional principles for defining the HMI architecture can also be formulated:

- Since monitoring is always active, it makes sense to have an overview display, or displays or zone(s) within displays, that supports this task.

- When moving to another display (e.g., a control display), reserve an area of that display to support the monitoring task.

- Monitoring tasks are strongly associated with alarm management (when exceptions or anomalies do occur). There is an abundance of guidance on how to take alarms into account in HMI design, but a detailed description is outside the scope of this book. However, at the HMI architecture stage, we should ensure that a few key attributes are reflected

Table 2–4 Key Characteristics of Generic Tasks

Generic Task	Key Characteristics
Monitoring	This task is always active. Therefore, the information groups required to sustain it must always be displayed, and for secondary information, available.
Controlling	This type of task usually occurs in a controlled fashion. Therefore, based on the results from the HTA, it is possible to design HMI elements that optimally support it.
Compensating	Ideally, pre-defined support for this type of task exists in the form of procedures that have been generated to deal with a set of abnormal, but expected, events. Executing procedures will often be done "through" the HMI, thus making it appealing to integrate the HMI and the procedures.
Diagnosing	Diagnosing must occur after monitoring has identified a threat to achieving one of the goals. Depending on how the information is organized, it is possible to trace a fairly direct path from a threat to its underlying cause(s).
Optimizing	This task can usually be carried out by using the same HMI components as those used to support "controlling" tasks.

in the characteristics of the alarms that will be presented to the operator through the HMI:

- The alarms should be useful and relevant, i.e., they should not be provided unless they provide information that will help the operator.

- They should have a defined response.

- The design of the alarm system must be such that the operator has enough time to respond to the alarm.

- And of course, the design of the alarm system should take into account the characteristics and limitations of the operator. This point will be addressed at the detailed design stage.

If additional guidance on alarm design is sought, then EEMUA 191 *Alarm Systems—A Guide to Design, Management and Procurement* should be considered. Other general principles for defining the HMI architecture include:

- Minimize the overall number of displays; if a given display requires minor (and this means *minor*) additions to support more than one task, then this should be done.

- A display must always contribute to achieving an operational (productivity or safety) goal; if a display cannot directly be linked to an operational goal, either the list of goals is incomplete or the display does not serve a useful purpose.

- The type of display to include in the architecture must be directly linked to the types of users identified during analysis of the context of use. For example, unless there is a compelling reason to do so, maintenance-related information should not be included in process management displays as this tends to make them more complex, requiring more training for the operators and increasing clutter.

- The number of monitors to use is normally extremely difficult to estimate. However, some fairly firm constraints can be formulated:

  - The absolute minimum for normal operation is obviously one monitor.

  - Commissioning frequently entails carrying out additional tasks that often require an additional monitor.

  - For redundancy reasons, it is often advisable to have available one more monitor than the absolute minimum required; this makes two monitors a recommended minimum in most situations where the proper operation of the process will require uninterrupted monitoring and control functionality.

  - Limited results from the literature have shown that operators infrequently consult a fourth (or more) monitor in petrochemical applications. If these results can be extrapolated to process control in general (admittedly a big "if"), then two to three monitors for a single operator should be about right.

One last point of caution is in order. There is no such absolute as the maximum number of loops an operator can effectively monitor or control; nor is there a maximum number of variables an operator can effectively track. If you remember our earlier discussion, lots of factors influence workload, making absolute predictions virtually hopeless. Further, monitoring a few poorly behaved variables may be all the operator can handle in a given situation, while monitoring a large number of well-behaved variables may be quite all right in another situation.

Using the above principles, it is possible to draw a hierarchy of the types of displays necessary to optimally, or quasi-optimally, support the user. It is also possible to allocate these types of displays to monitors, or at least to zones on monitors.

Once this first architecture is defined, it can be validated. Given that the most costly mistakes are made early in a project (Maier & Rechtin, 2002) it is particularly important to have a sound architecture before work proceeds to the next phase. Scenarios based on the HTA should be generated, and the flow of the user's navigation can be examined on a paper representation of this architecture. During this validation, it is important to ensure that the architecture includes all of the displays required to carry out the scenarios. The validation can also be used to identify opportunities to simplify the user's interaction with the HMI. Scenarios should be chosen to represent the important tasks of the user and since this table-top validation exercise is extremely inexpensive, this is an excellent occasion to enhance the quality of the resulting HMI and system. Once this validation has been completed, it is also a good idea to capture the resulting HMI architecture and to obtain some buy-in from project management. Needless to say, it is also good to be prepared to make some adjustments to it as we learn more. Here again, there are no hard-and-fast rules on how much time will go into creating the architecture of an HMI. However, the author's personal experience indicates this is normally a matter of days, rather than a matter of weeks.

We emphasize the notion of architecture for several reasons, one of which is that once a sound HMI architecture has been defined and validated, making the step to detailed design is extremely simple. In that case, detailed design consists of creating an instance of each type of display by relying on a set of detailed design rules. Figure 2–6 illustrates this approach.

The top half of Figure 2–6 is the overall architecture for a fictitious computer-based HMI. The architecture consists of a monitoring window, a control window, a set of diagnosis windows, and a window supporting ancillary tasks. Note that the content of each of those architecture-level windows is composed of placeholders for components that will be specified at the detailed design step. Given that architecture, this detailed design step becomes extremely easy and consists of identifying and applying detailed design rules for the components (these design rules can be found in the appendix). For many components, specific detailed-design patterns exist; in case pre-defined patterns do not exist, then individual design rules may be used to design the appropriate components. Aside from these detailed design rules, there are some simple, higher-level principles that have proven their worth:

- Maintain visual momentum. This academic-sounding principle is in fact very simple and refers to the fact that, at any given time, the user needs to know where he or she is coming from in the HMI, where he or she is now, and where he or she can (and should) go. There are many ways to promote visual momentum. The best way is to ensure a good fit between the HMI architecture and the user's task (here again, the HTA comes in handy). Another way to ensure consistency between components of the HMI is to enforce consistency in a number of areas such as the location of display elements, font sizes and style, color, labeling, etc.

- Aim for simplicity, both real and perceived, when designing the HMI. Real simplicity again comes into play when the content of the HMI optimally matches the information and control requirements identified through the HTA. Perceived complexity can be helped by a few simple rules. Consider the following (Figure 2–7).

Both windows contain exactly the same number of items; yet, which one appears simpler? The reason for the apparent reduction in complexity is the reduced number of vertical alignments necessary to line up the elements. Another useful principle to the same effect is known as *reduce the ink*. Simply put (no pun intended), this leads to

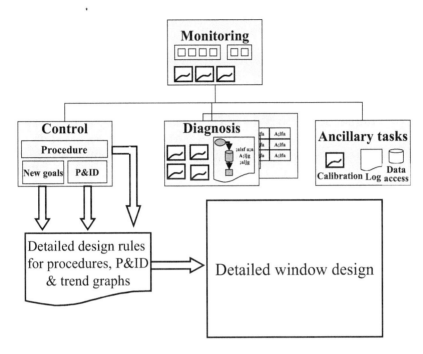

Figure 2–6 *Example of Transition from HMI Architecture to Detailed Design*

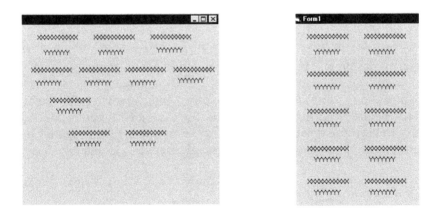

Figure 2–7 *Example of Reduction in Perceived Complexity*

using the minimum number of visual symbols (lines, characters, etc.) required to transmit the information.

- ■ Error prevention and recovery. Here again, this principle argues that it is better to prevent errors, and, if they occur, recovery must be simple. There are a number of detailed implementation techniques that can be brought to bear. For example, consistency in the use of acronyms, local validation of inputs, as well as the use of masks (e.g., "mm-dd-yy" for indicating the correct format for a date entry) usually yields good results.

- ■ The use of human visual pattern recognition to allow decisions to be made by recognition rather than by analysis is also good practice. Figure 2–8 shows a simple example for circular analog indicators.

- ■ As Figure 2–8 shows, when indicating the nominal or "normal" value, the arrows should point to a normalized 12 o'clock position. The task of monitoring this HMI component thus becomes "ensure that the arrows are pointing straight up" rather than "ensure that each of the indicated values is within the accepted value range." This approach reduces the user's workload and will lead to faster monitoring and fewer human errors.

It is important to note that for reasons of cost and efficiency, the detailed design will be produced on a computer or paper mock-up.

All of the preceding steps may make sense, but the saying "the devil is in the details" definitely holds true in the case of HMI detailed design. To help keep the devil at bay, the next section will go through a small but detailed description of architecture definition and detailed design.

Figure 2–8 *Example of Using Human Pattern Recognition*

Example: Producing Design Solutions

To give an example of how an HMI is designed, let us carry out the step-by-step design of a first window, based on the example we have used so far. Let us assume that the overall HMI architecture shown in Appendix A fits our purpose. We will now design the monitoring display for this architecture.

1. The first step is to use an appropriate HMI template. This template will normally have been specified during the HMI architecture phase. For the purposes of this example, assume Figure 2-9 is a suitable template.

   We have chosen to use the style guide from a popular window-based operating system, and this has helped us determine the nature and content of the menu items that appear at the top of the display. The row of buttons underneath the menu items is a standard toolbar containing buttons that provide shortcuts to elements from the menu; here again, the style guide provided valuable guidance. The template also makes provisions for a title for each individual window. This will be useful should software troubleshooting be necessary. The main zone available for information is thus the Work Zone. This is where we can put material that will support the operator's tasks.

2. Once we have picked the template, we need to determine the display's content. Common human-factors wisdom is that the results from the HTA should dictate the content. This is good guidance, but additional considerations should be factored in, such as:

   2.1 Monitoring displays should support the "management-by-exception" and "management-by-awareness" behaviors.

   2.2 The HMI should contain information allowing the operator to determine if the set of operating goals (usually productivity and safety) supporting the mission of the system is being met. In our case, there is good agreement between these operating goals and the results from the task analysis that we completed earlier (see Figure 2-5).

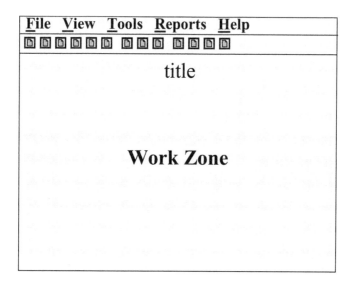

Figure 2-9 *Example Template for Display*

In our case, the production goals are to produce paper of the right basis weight and moisture content (respectively, 48 grams per square meter [gsm], and 8% moisture), while the quality goal is to ensure that the profile of variation of various measured variables remains within a given tolerance band. Our safety goal is to ensure that effluents do not exceed a certain threshold. These goals correspond closely to those identified during our HTA (a fragment of the HTA for monitoring tasks is reproduced in Figure 2-10).

With this information, we decide to use simple numerical indicators that will display the current basis weight and moisture content of the sheet. For the quality information things are a bit touchier, since there are several variables for which profiles have to be within specifications. For the time being, we plan to use a single "good/bad" indicator for the various profiles. For effluents we have a single measurement, but plan to use the same type of indicator as for the profile information. All of the previous indicators will use similar visual coding: When the parameters are within tolerance, the background of the indicator will be color coded in green; an unfavorable trend or marginal value will color code the indicator's background in yellow. An off-spec condition will yield a red background with a flashing border[4].

Producing Design Solutions   51

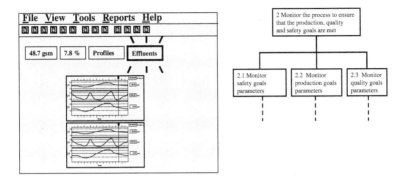

Figure 2–10 *Sample Monitoring Display and HTA Fragment*

We just took care of the "management by exception" business. Now, we want to provide information to help the operator manage by awareness. We remember that operators rely on monitoring key parameters, and that trend graphs are often used for this purpose. In our case, the information collected during the HTA suggests that those key parameters closely match those that can be used to ascertain whether the operating goals are met. We then decide to make trend graphs available for all of the variables that support a given operating goal[5]. For example, in the case of basis weight, the variables that "drive" basis weight are stock valve opening, machine speed, and stock consistency. Figure 2–10 shows the resulting display.

While this display seems quite logical and well-aligned with the HTA and other design requirements that have been identified until now, we resist the temptation to declare the job done. Rather, we realize that once we have a few more displays complete, we will need to perform usability testing to ensure we understood the problem right and that our brilliant solution (on paper) remains brilliant in the field. Usability testing is the topic of the next chapter.

---

4. Datasheets with recommended guidelines for designing numerical and binary "good/bad" indicators are provided in the Appendix.
5. A datasheet with recommended guidelines for designing trend graph is provided in the Appendix.

Specification of the Resulting HMI

Depending on the needs of the project, the specification phase can be more or less formal:

- In some cases, the mock-up and the documents produced are used as the HMI specifications.

- In other cases, a formal specification document, such as a system or software requirements specification, may be required.

It is very important to make this choice early during the project. It will thus be possible for the HMI design team to produce the kind of specification required at the best cost. Experience suggests that:

- A mock-up is useful to avoid ambiguities or to clarify various aspects of a specification.

- Good communication between the HMI design and specification team and the development team is essential. Good communication can be enhanced by good interpersonal relationships and by physical proximity of the teams.

There are many formats that can be used to transmit design requirements to the implementation team. However, in the author's experience, it is often useful to break the HMI's contribution to the overall specification into a *general part* and a *specific part*. The general part usually addresses permissible response delays, visual coding of the information, character sizes, navigation mechanisms and navigation map, style guide, alarm handling, hardware (displays and pointing and data entry devices), and templates. The specific part typically describes individual panels or displays and often includes a pictorial description of the item and a description of the specific controls and display elements it contains.

An HMI specification document, which can include the mock-up, is normally produced at the end of this phase. This document must be reviewed, approved, and issued before the HMI design phase can be considered complete. If additions or modifications to the HMI design are required, the specification must be updated, reviewed, approved, and re-issued.

Integrating Software-based and Hardwired HMIs

Integrating software-based HMI components and hardwired panels is a common issue. Obviously, there are many more degrees of freedom in designing software-based HMIs than hardwired panels, sometimes to the point of arguing that designing a bad hardwired panel is difficult (the author has a few telling stories of how bad hardwired panels can still be designed).

It is customary in many industries that software-based HMIs are used for process optimization while basic functionality is provided by hardwired panels. In some instances, solely software or hardwired HMI designs are used. Nevertheless, in most situations, both approaches are likely to cohabit for a long period of time, and this calls for some serious efforts at making those components work together as seamlessly as possible.

Here again, a few down-to-earth principles can be stated:

- There should be a convention that determines which functionality will be allocated to the software and which to the hardwired components. For example, safety-related functionality could be allocated to hardwired panels, and process control and optimization could be allocated to software-based HMIs. What is important is that this convention be applied consistently (remember the discussion on visual momentum).

- The same visual coding conventions (e.g., color, fonts, labeling) should be used for both technologies.

- The same material, energy, and information flow conventions should be used for both technologies.

- The same alarm handling strategies should be used if alarm handling is provided for both types of HMI technologies.

- While HMI architecture is being created, the designer should try to minimize the number of occasions when the user has to move between the software- and hardwired-based HMI components to carry out a task.

Through those simple rules, integrating various HMI technologies can be simplified and the overall performance of the user-HMI-application triad optimized.

Transition

In this chapter, the various steps leading from the analysis to the design of HMIs, be they software- or hardwired-based, have been explained. Normally, evaluation of the resulting HMIs should also have been included. However, since HMI evaluation can also be carried out on an existing system, it is convenient to treat it separately for readers mostly interested in the evaluation of existing HMIs. It is important to realize that this discontinuity in the treatment of the subject matter is somewhat artificial and that HMI evaluation is very closely related to HMI design.

References and Additional Readings

The following are the references as well as additional sources of information readers may find useful. Some are very design-oriented, while others explain the underlying philosophy of optimizing the user-HMI-application triad.

Paper-based Publications

Bainbridge, L., "Ironies of Automation, Automatica," Vol. 19, No. 6, pp. 775–779, 1983. (A very good summary of the challenges facing the ever-growing spread of automation in complex systems; still very relevant.)

Conant, R.C., Ashby, W.R., "Every good regulator of a system must be a model of that system," in the *International Journal of System Sciences*, 1970, Vol.1, No. 2, pp. 89–97. (This paper provides an indirect, and more technical justification, for the notion of mental models.)

Cooling, J.E., "First steps - requirements analysis and specification," in *Software Design for Real-Time Systems*, 1991.

Fiset, J.-Y., "Conception d'interfaces humains-machines pour la conduite de systèmes complexes," Unpublished Ph.D. dissertation, École Polytechnique de Montréal, 2001.

Kim, I.S., Modarres, M., "Application of Goal Tree-Success Tree Model as the Knowledge-Base of Operator Advisory Systems", *Nuclear Engineering and Design*, 104 (1987), pp. 67–81. (A good introduction to functional modeling.)

Kirwan, B. and Ainsworth, L.K. (Eds), *A Guide to Task Analysis*, Taylor and Francis, London, 1992, 417 p.

Maier, M., Rechtin, E., *The Art of Systems Architecting*, ISBN: 0-8493-0440-7 (Hardbound), CRC Press, 2002.

Moray, N., Lootsteen, P., Pajak, J., "Acquisition of Process Control Skills", IEEE Transactions on Systems, Man, and Cybernetics, Vol. SMC-16, no. 4, July/August 1986, pp. 497–504. (This paper provides a very good discussion of how operators learn to understand and control a system.)

Potter, S. S., Woods, D. D., Hill, T., Boyer, R. L., Morris, W. S., "Visualization of Dynamic Processes: Function-Based Displays for Human-Intelligent Systems Interaction", Proceedings of the 1992 IEEE International Conference on Systems, Man, and Cybernetics, Chicago, IL, October 1992. (An interesting description of how functional modeling can be used to design HMIs.)

Rasmussen, J., "Skills, Rules, and Knowledge; Signals, Signs, and Symbols, and Other Distinctions in Human Performance Models", IEEE Transactions on Systems, Man, and Cybernetics, Vol. SMC-13, May/June 1983, pp. 257–266.

Rasmussen, J., "The Role of Hierarchical Knowledge Representation in Decision-Making and System Management", IEEE Transactions on Systems, Man, and Cybernetics, vol. SMC-15, no. 2, March/April 1985. (An interesting, albeit somewhat abstract, discussion of how humans understand a system.)

Reason, J., *Human Error*. Cambridge, U.K.: Cambridge University Press, 1990.

Zwaga, H.J.G., Hoonhout, H.C.M., "Supervisory control behaviour and the implementation of alarms in process control", chap. 7 in Human Factors in Alarm Design, Stanton, N. (Ed.), Taylor & Francis, 1994.

Web-based Resources

The following is a short list of non-commercial Web sites (well, OK, there is a commercial one, but it is not related to the author and it contains quite a bit of useful material) that are relevant for HMI design for complex systems.

http://www.asmconsortium.com/asm/dashboard.nsf?Open, the site of the Abnormal Situation Consortium, grouping a number of partners in an effort to better manage process upsets.

http://www.enre.umd.edu/ifmaa/index.html, the site of the Functional Modeling group, which is interested in better understanding goal-oriented representations of industrial and other processes. Many of the ideas discussed in this book, about using goals to organize the information in the HMI, are strongly connected to functional modeling.

http://www.hse.gov.uk/, the site of the British Health and Safety Executive, a great source of information on HMIs, alarm handling, process safety and several other aspects of human factors engineering.

3

Evaluating the Design Against the Requirements

This chapter describes techniques to carry out the evaluation of HMIs, whether they are part of an existing system or are in the process of being designed. A point worth mentioning is that learning how to evaluate HMIs is a very good training ground for designers, as few people will knowingly re-create mistakes.

Description of the Evaluation of Design Solutions

This section describes two well-known techniques for evaluating HMIs. Each technique has advantages and shortcomings, yet their combination is very potent.

Heuristic Evaluations

The first method consists of having a small group of experts inspect portions of an HMI to determine its conformity with a set of heuristics (*heuristics* refers here to a set of well-grounded design principles or good practices). In another context, such as the Web, this has sometimes been referred to as "discount usability engineering." This could in fact be the case, but with a principled approach and sound knowledge, heuristic evaluations offer a powerful tool to discover and remedy severe usability issues before the HMI gets deployed. Additionally, mastering the fundamentals of heuristic evaluation helps the designer avoid specific design errors.

Heuristic evaluations are a type of non-performance-based (or *static*) test; that is, the objective is to ensure that the HMI conforms to generally accepted "good practices." There is, however, no attempt to ensure that the HMI is optimally suited to supporting the task at hand. This latter aspect may be addressed by another technique known as *usability testing*, which will be introduced shortly. The advantages of heuristic evaluations are:

- They can be carried out rapidly—a display or panel can typically be evaluated in a few minutes.

- They are relatively inexpensive.

- They can identify major usability problems.

- Best of all, they do not require user participation. Note that this does not imply that user inputs will not be considered in the project; in fact, users will be involved in a complementary evaluation (i.e., usability testing). Rather, heuristic evaluations aim at identifying discrepancies between the candidate design and good design practices.

Unfortunately, heuristic evaluations also have a few shortcomings:

- The technique looks and sounds deceptively simple but it requires substantial expertise in "good practices" for HMI design.

- Generally, it finds only 30% to 50% of problems affecting users.

- The main shortcoming, though, is that by definition, heuristic evaluation will not, and cannot, identify a lack of task support for a given HMI. The reason, of course, is that the user's task is not taken into account.

In spite of those limitations, heuristic evaluation is the method of choice when time is severely limited or when it is desired to get a good first idea of the situation with an existing HMI. Its usefulness is somewhat more limited if the designer is knowledgeable about the heuristics and has already taken them into account during the design phase; this is, however, quite rare in practice. In that case, heuristic evaluation will mostly be used as a method for confirmation or verification. The procedure for carrying out a heuristic evaluation is very simple:

1. Inspectors are selected. The qualification required is familiarity with the heuristic evaluation process. It is helpful, but not indispensable, for the inspectors to be knowledgeable about the domain. Remember that the purpose of heuristic evaluation is to ensure consistency between the design and the heuristics, not to ensure that the content is adequate for the job at hand. Normally, three to five inspectors will be used. However, depending on the qualifications of the inspectors, this number may vary somewhat.

2. The inspectors familiarize themselves with the HMI. This usually requires some coaching and either a paper or electronic mock-up, or even the real system if an existing HMI is being evaluated. It is also possible in many cases to rely on a paper document containing screen shots of the HMI (e.g., a design or user manual).

3. Each inspector evaluates the HMI independently with respect to a list of heuristics, although the author's experience suggests that pairs work well, too. Individual results are logged. Table 3-1 provides the list of heuristics commonly used (Nielsen & Mack, 1994).

4. Results from the individual inspectors' results are consolidated and ranked in terms of their significance. Often, a three-level ranking (e.g., major, medium, minor) is adequate.

5. Recommendations for improving the situation are usually provided. The recommendations should be limited to the areas for which heuristic evaluation is suitable. Inspectors need to remember they do not have sufficient information on the user's task.

This procedure for heuristic evaluation is the one commonly used. However, extensions to the basic method have been found to work well, in the author's experience. For instance, before the heuristic evaluation, it often pays handsomely to draw a navigation map. This will help in understanding how the user has to go through the displays of the HMI to carry out his or her task, and will assist in picking

Table 3-1 List of Heuristics

Heuristics	Comments
Visibility of system status	At any time, the user should know what the system is doing. For example, a progress bar provides indication that an action is being processed, or a visible change on the display indicates that the system has changed state based on user input.
Match between system and user's world	The system "speaks" the user's language rather than in cryptic codes. For example, not all users are fluent in Grafcet (a language used by control and automation engineers), yet many designers wish to explain the control logic to the users in this language.
User control and freedom	The user controls the HMI, and not the other way around.
Consistency and standards	There should be consistency between the various elements of the HMI. For example, software-based HMIs should conform to the style guide for the platform.
Error prevention	The design should attempt to prevent errors. For example, confirmation windows and legible labels help prevent errors.
Recognition rather than recall or analysis	Repeat the required information if it is needed at more than one place on the HMI. Also, use pattern recognition rather than detailed dial or gauge analysis.
Flexibility and efficiency of use	To accommodate various skill levels, there should be more than one way to do things.
Aesthetic and minimalist design	The resulting HMI design should be as simple as possible, thus elegant.
Help and documentation	It should be simple and short, adapted to the task and to user characteristics.
Help users recognize, diagnose and recover from errors	No matter how many efforts are made, errors will occur. The HMI should thus help users recover from them. For example, an alarm should be linked to an alarm sheet that indicates the nature of the alarm, its likely causes and the recommended path to resolve it.

up inconsistencies or anomalies in the navigation. It will also help comment on strategic, rather than tactical, issues with the HMI. Also, while carrying out the heuristic inspection, it is important to be on the lookout for the use or misuse of the HMI design principles presented previously.

You should use heuristic evaluation when the detailed design is fairly stable; however, it can also be used right before the deployment of a new system (to see if there is danger of major flaws in the system). Finally, it can be used as a diagnosis technique when issues arise during or after deployment.

Finally, it is important to realize the potentially powerful effect of heuristic evaluations. The purpose of these inspections is to identify deficiencies…and that's what they do. Deficiencies, in turn, are often perceived very negatively. Proper planning from the start on how, when, and to whom results will be communicated will ease their acceptance.

Example of Heuristic Evaluation

The following example, Figure 3–1, will show how even a simple HMI component can be problematic.

Normally, one of four answers can be provided to an individual heuristic: pass, fail, cannot say, or not applicable. The heuristics and the rating that was given to each follow, with a short explanatory comment:

- Visibility of system status—not applicable.

- Match between system and user's world—pass. There is no unambiguous language on the window.

- User control and freedom—pass. The user can close the window when he or she so desires.

- Consistency and standards—fail. The greyed-out closing window control does not conform to the style guide. The accelerator or shortcut (the underlined N of the No button) is not applied consistently. The Y of the Yes button should also be underlined.

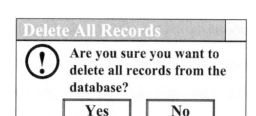

Figure 3–1 *Sample Window for Heuristic Evaluation*

- Error prevention—fail. No button has a default focus. Pressing "Enter" when this window appears will have unpredictable results.

- Recognition rather than recall or analysis—not applicable.

- Flexibility and efficiency of use—fail. The shortcuts should allow flexibility and efficiency, but they are not used consistently.

- Aesthetic and minimalist design—pass.

- Help and documentation—not applicable.

- Help users recognize, diagnose and recover from errors—not applicable.

This example shows that even simple HMI components may have several problems, with some having the potential to lead to very serious consequences. The example also shows why designers should design, rather than users. Including default focus on a button and implementing shortcuts are normally part of the designer's set of concerns, rather than the user's.

Heuristic evaluation also provides valuable clues as to the expected state of the HMI. If several issues appear during this evaluation, one can expect other problems with the adequacy of the HMI for supporting the task.

Usability Testing

Usability testing provides the "other side of the coin" for evaluating the HMI. It consists of having representative users simulate a selection of their tasks by using a mock-up of the intended HMI. Usability measurements are then taken. Remember that we define usability as "the extent to which a product can be used by specified users to achieve specified goals with effectiveness, efficiency and satisfaction in a specified context of use." Usability tests are sometimes referred to as performance-based tests, since it is possible, and indeed desirable, to measure performance objectives.

A key difference between heuristic evaluation and usability tests is that the latter are an indispensable part of the HMI design process. Following a first mock-up, usability tests will be performed to ensure that the resulting design meets user and organizational requirements. Usability tests offer the following advantages:

- They can identify, a priori, serious or recurring usability problems before a single panel gets built or before a single line of code is written. As such, they contribute very effectively to reducing the technical, schedule-related, and financial risks of a project.

- They help focus efforts on serious issues rather than on less-significant issues.

Usability tests also have a few disadvantages:

- In spite of the relative simplicity of the procedure, it is important that the staff conducting usability testing be well-qualified.

- There is large variability in the way usability tests are conducted.

- Usability tests are relatively expensive in terms of resources (including users), time and money. They are, however, still far cheaper than lost production, safety threats, unsatisfied customers, a poor reputation or software redesign.

The procedure for carrying out a usability test is actually very simple:

1. Define test objectives. Those objectives will vary depending on the time at which the test will be carried out. If usability testing is part of the HMI design cycle, then the objective of the first test is often simply to successfully carry out the simulated task. On a second round of usability tests, consideration will be given to efficiency and criteria such as number errors, user satisfaction, time to carry out the task, time to learn, etc. Eventually, test objectives dealing with user and organizational requirements will have to support the mission and other objectives set out during the analysis phase.

2. Identify tasks, taken from the task model, that will be tested according to frequency, criticality, complexity, etc., and generate short narratives for these tasks.

3. Secure a mock-up of the HMI. During the initial tests, we advocate a paper-based mock-up, although it can be created using a computer and then printed out. During subsequent phases, if ease of navigation becomes an issue for which usability testing is desired, an electronic mock-up with minimum functionality could be used.

4. Once you feel ready with your objectives, narratives and mock-up, carry out a pre-test with one member of your staff to validate the mock-up and test scenarios. This step aims at ensuring that you will not waste valuable user time. It has saved the author immeasurable embarrassment by allowing mistakes or other ambiguities to be removed prior to the actual usability tests.

5. Select and solicit representative users of the system. Representative users will normally heavily include operators (since they are the primary users) and may also include maintenance or technical users (if they are expected to also be significant users). Avoid recruiting friends as this may hamper their (or your) evaluation of the system. The question of how many users will participate is difficult, but not impossible, to answer. Rules of thumb are:

    - Since the initial round of usability tests simply aims at determining if the design "can do the job," typically, two to three users will suffice. For each of the subsequent rounds, very limited research indicates that seven or eight users may uncover most of the usability problems. In industrial applications, prospective users may be in short supply, so it is probably safe to go down as low as five users per round.

- Do not involve the same users on subsequent rounds as they will likely remember their previous rounds, and this may skew the results.

6. Prepare required equipment (e.g., video camera, confidentiality agreement).

7. Determine with client or management, prior to testing, how confidential material (e.g., video tapes, results) will be handled.

8. Carry out the actual tests. It is important to realize that usability testing for an HMI is done with individual users, rather than with a group of users. For each selected user in a round of tests:

- Greet the participant.

- Explain test objectives, with particular emphasis on the fact that it is the HMI that is being tested, not the participant. Address any need to assure the participants that the results will be kept confidential and that only aggregate data will be used. Discuss "thinking aloud" during the test, and how the information collected during the test will be used.

- Ensure that all questions from the participant are answered and that he or she is comfortable with carrying out the tests.

- Have the participant carry out the scenario.

- Only interrupt as needed; if the participant is stuck, encourage him or her, and provide cues with moderation.

- Maintain a positive attitude irrespective of the results. After the test, debrief the participant to identify positive points and those to improve.

9. If required, have the participant complete an evaluation questionnaire. This questionnaire could include an overall evaluation (you may want to ask for a rating from 1 to 5), a short description of the improvements suggested by the user, a short description of the strong point(s) of the current design (if any), as well as any other suggestion that the user may have.

10. Analyze and consolidate results.

11. If traceability of HMI design changes is paramount, then consider preparing a comprehensive usability report containing the results. From there, make recommendations and then improve the mock-up. On the other hand, if efficiency is paramount, file the current paper prototype as well as notes and the recordings (if any) from the usability tests, then improve the mock-up without preparing a written report. In the rare cases where you elect to produce a comprehensive usability report, the Common Industry Format provides a complete template[1]. In this author's opinion, the best way to use this template is to adapt it to one's particular situation rather than to try to apply it blindly.

12. Until the tests reveal that no major improvement of the mock-up is required, go back to step 1 of this procedure. Expect that, if the analysis went well, there should be between three and five (in most cases) rounds of usability tests.

For usability tests to run smoothly, there are a few tricks of the trade to know. First, it is important to strictly limit the number of staff involved. Normally, there will be the moderator (you) who is running the test, with perhaps one assistant to control the mock-up (if a paper mock-up is used). If the scenario or the mock-up is complex, you may want to ask for one observer to take notes. A better way, in the author's experience, is to use a small MP3 voice recorder. These devices are inexpensive, non-intrusive, and with the user's permission, make it easy to keep a digital record of the test results.

---

1. ISO/IEC 25062:2006 Software engineering—Software product Quality Requirements and Evaluation (SQuaRE)—Common Industry Format (CIF) for usability test reports.

It is also important to minimize the number of staff interacting with the users; normally, only the moderator will interact with the participant. Another good idea is to remember that usability tests give a good idea of what works and does not work in a design, but they are not a rigorous laboratory evaluation, so some discretion and judgment on the designer's part is required. Finally, wherever the designer can, he or she should take advantage of existing knowledge rather than rediscover it through usability testing. For example, numerous existing standards provide virtually all the information required to ensure the legibility of labels. It is particularly inefficient to attempt to determine if information is legible through usability testing.

The following example will illustrate how to carry out a usability test.

Usability Testing Example

We will now go through the motions of planning and carrying out a usability test. Since we have done the heuristic evaluation example on a software-based HMI component, here we will describe usability testing for a hardwired panel design used for a fictitious task (here, to establish flow through some piping). Taking it step by step:

- We just designed a small hardwired panel (Figure 3–2) and want to test it before we actually get it built. Since the design is simple, we do not foresee any specific issue with it. The objective for our first test is to ensure that the user can establish flow without serious problems.

- We produce a short (here, four to five lines) narrative to explain the task to the user, and ask him or her to carry it out. The narrative is just a guide, and since we will talk it through, the exact verbiage is not critical.

- Being astute designers, we use a piece of cardboard on which we hand-draw the elements of the panel.

- We feel quite sure of ourselves, but having just re-read the chapter on the evaluation of the design against the requirements, we ask a colleague to act as a user and to go through the motions with our design. This takes only twenty minutes, but allows us to realize that it would be a good idea to label the panel mock-up in the same way as the panels are

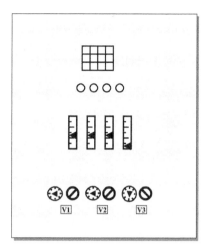

Figure 3–2 *Sample Hardwired Panel*

labelled in the control room. We now feel ready to try it out with real users.

- We take care to remove mistakes or other ambiguities prior to the tests.

- We ask the superintendent for permission to carry out a first round of tests. She agrees, provided the operator on shift also agrees and we do not interfere with operations. We assure the superintendent that we will pick a time when things are going smoothly, and that we will not interfere with the operator (besides, we are sure that the operator will tell us if we do). We also inform the superintendent (and our own management) that the results from an individual will not be disclosed but that only the aggregated results will be disclosed. This is agreed to after some explanations are given to the superintendent (failure to agree would have prevented us from carrying out the test).

- We contact the operator and explain what we would like to do and why. He enthusiastically agrees, and we set an appointment on the following day.

- At the agreed-upon time, we show up at the control room and ask if the situation is still favorable for the test. It is, so we propose to get started. We then explain the test objectives, with particular emphasis on the fact that it is the HMI that is being tested, not the operator, and we explain that the results from this test will be kept confidential. The operator has no special questions, but is quite curious to see if issues he mentioned while we performed the HTA have been addressed. We confidently assure him that they have.

- Having confirmed that if the process requires his attention, he will tell us and will go back to look after it, we show him the mock-up of the panel and ask him to carry out the task described in the narrative. This consists of:

  - Start from a given configuration (Figure 3–3) and establish flow through a pipe. The starting configuration shows that flows already exist in the first three pipes. The operator must turn on the pump (with the on-off switch) and then set its speed.

  - To establish flow, the operator sets the pump speed to a value similar to that of the neighboring controls. Since the pump has not been turned on, an alarm light comes on (see ② on Figure 3–4) and the flow remains at 0. We indicate these results to the operator on the mock-up.

  - We then ask the operator to make another attempt. After closely examining the settings of the adjacent controls, he then turns on the pump and adjusts the pump speed. We then show him the second cardboard mock-up with the flow having been established (see Figure 3–5).

- At this stage, we have learned:

  - There is indeed sufficient functionality on the panel to establish flow (admittedly, this is a simplistic scenario).

  - The fact that the pump must be turned on prior to adjusting its speed may not be obvious enough. We then decide to carry out the same test with two other operators.

- Being the astute designers that we are, we start wondering whether the fact that the pump speed control knob is to the left of the pump on-off switch might not be problematic. To determine whether that is the case, we quickly modify the

70 Chapter 3—Evaluating the Design Against the Requirements

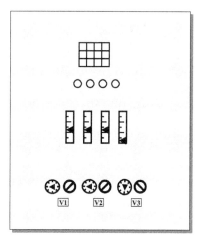

Figure 3–3 *Mock-up 1*

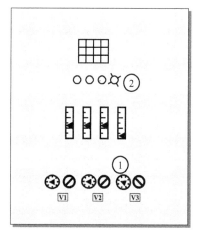

Figure 3–4 *Mock-up 1 with Indication of Results*

design (a pencil and eraser work great) and decide to re-test with two other operators.

The above example is obviously simplistic but it has actually been used in a similar application (the production of electricity by a water turbine-driven alternator) with a fair bit of success.

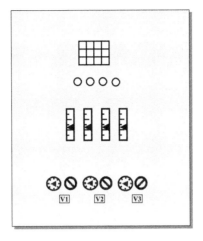

Figure 3–5 *Mock-up 2*

Transition

In this chapter, we learned how we can evaluate whether an HMI design conforms to good design practices and meets its intended purposes. Once we have determined it does, our next step is to capture the design into a set of requirements that will be included in the specification of the new system. This will be the subject of the next chapter.

References and Additional Readings

The following are references as well as additional sources of information readers may find useful.

Paper-based Publications

Nielsen, J., Mack, R.L., "Usability Inspection Methods," John Wiley and Sons, Inc. 1994. (An interesting book that deals in some detail with heuristic evaluation, as part of a family of techniques known as *discount usability testing* methods.)

Web-based Resources

http://software.gsfc.nasa.gov/AssetsApproved/PA2.5.1.3.pdf, a handbook on usability testing from NASA's Goddard Space Flight Center. Nicely done, with many useful and practical tips.

4
Specifying an HMI

As previously mentioned, the output of HMI design work will be a contribution to the *system requirements specification* (SyRS). This is usually no trivial matter for several reasons. Frequently, the SyRS (or any other statement of system requirements) simply does not exist as such, and the requirements are spread among various pieces of documentation, or worse, are in the collective minds of the project sponsors and team. When it does exist, the statement of system requirements is often incomplete or lacks rigor. This situation is compounded by the fact that it is difficult to write good requirements.

While it is beyond the scope of this chapter to offer comprehensive coverage of requirements writing, we will look at a few principles and rules that will help in producing a satisfactory statement of HMI requirements. Success in doing this will considerably improve the odds of properly building the HMI that emerged from the previously described method. Further, experience has shown the HMI specification constitutes the first cohesive description of the system's functionality. For that reason, it is often used as the embryo of a more comprehensive statement of system requirements. Before we get started, and paraphrasing an industry standard (IEEE 610), we will ensure that we are on the same page with respect to two important definitions:

- A requirement is a condition or capability needed by a user to solve a problem or achieve an objective, or that must be met or possessed to satisfy a contract, standard, specification, or other formally imposed document. Note that a representation (e.g., text, graphic) of a requirement is also called a requirement.

- A requirements specification is a document specifying the requirements for a system or component.

We will start our exploration of how to specify an HMI by looking at some relevant standards. Then we will examine how to actually specify various HMI elements and we will illustrate the process with an example.

Standards and Guidelines

There are many sources of standards that can be used to write requirements, but none that deal specifically with HMIs. This is not really a problem, as requirements (be they related to HMIs or to other system or software components) share characteristics across domains. There are several good sources to draw upon for producing specifications; two such sources can be found in software (IEEE 830) and system engineering (IEEE 1233). We will briefly touch upon each.

IEEE 830 is surprisingly readable (for a technical publication) and offers good advice on how to write specifications and requirements. It also specifically addresses HMI-related requirements and goes as far as to consider what the specification should cover: "...logical characteristics of each interface between the software product and its users...e.g., required screen formats, page or window layouts, content of any reports or menus or availability of programmable function keys...." In addition, the following points are based on an interpretation of elements of this source, and are especially relevant to the specification of HMIs:

- A requirement should be traceable. For HMIs, this implies that every requirement must contribute to achieving the system's mission, rather than reflect something that gets added later in the process because, for example, it would be nice to have.

- Requirements should be ranked for importance or stability. As we have seen previously, HMI design is iterative in nature. There may be situations where a given aspect will only be determined when a system has been developed further. For example, data availability may be influenced by which specific model of sensors is installed, and bandwidth issues may impact overall system response time. It is important to recognize these issues and to ensure that appropriate leeway is built into the requirements and specification.

- A specification should be verifiable. That is, it should be possible to test all of its requirements. This is an especially useful characteristic as it keeps stakeholders and designers honest about the HMI's design. For example, statements to the effect that "the HMI shall be easy to use" or "navigation shall be efficient" are inherently untestable and will be kept at bay if one insists on verifying the resulting specification or requirements.

IEEE 830 deals mostly with software-based documents; IEEE 1233 addresses system needs (broadly speaking, encompassing both software and hardware). While both standards are essentially in agreement over the nature of requirements, the latter goes into somewhat more detail regarding the context of system use, potential sources for requirements, and the process one could use to obtain requirements. As with IEEE 830, this standard addresses the specification requirements related to HMIs, as well as the inclusion of many pieces of information captured during the analysis portion of the HMI analysis-and-design process described previously. Given the importance of properly defined requirements to enhance quality, reduce costs, and ensure that we "build the right thing" and that we "build it right," the reader is encouraged to consult these documents.

How to Specify an HMI

Having acquired some knowledge on requirements and specifications, we can now examine how we will actually specify the HMI. Doing this will be helped considerably by taking advantage of a few rules.

Ground Rules

The first rule proposed is that one must consider the needs of specification users. This might seem a strange statement, but it is important to recognize that being overly detailed on the specification may add costs without adding anything to the design (e.g., specifying that a symbol is to be displayed in yellow may be sufficient, without spelling out its red-green-blue components). Specification users can be of different types. For example, software developers will usually use the specification to develop the software (or one would hope that they would!). However, other stakeholders may also have an interest in the specification. For example, user representatives are often asked to sign-off on the specification as a statement that what is described meets their needs. It is important to strike a balance among the level of formality and details of a specification, its understandability and its cost. Here are a few tricks to help find a suitable balance:

- Good relationships and close communications between those responsible for producing the specification and those responsible for implementing it will go a long way toward producing a document that captures the essentials without getting bogged down in trivia.

- Including screen captures or another graphic representation of the last version of the mock-up that emerged from the usability tests (see previous chapter) will help to simplify the specification and keep it readable.

- The proper use of a good requirements writing style will help decrease ambiguity and will ease communication among the parties involved.

Obviously, if the physical distance among the parties is large—for example, with off-shore development—the specification may have to be more detailed. However, one should always keep in mind that the specification is a means to an end, rather than the end itself. Modern Internet-based communications, such as e-mail and video-based teleconferencing, often permit keeping the specification reasonably straightforward if frequent and short communications occur.

Another rule is to determine very early how HMI requirements will be captured in the statement of system requirements. As we mentioned before, it often occurs that the HMI specification forms the embryo of the system requirements. However, if a defined process and associated documentation structure already exists, it is important to determine how HMI requirements will integrate with those, how versioning will be accomplished, etc.

Finally, pay particular attention to making the specification directly usable by the developers. For example, avoid such statements as "the font should subtend a visual arc of so-and-so minutes of angle;" instead, state the resulting physical value.

Starting to Specify: Introduction

Even though the specification is intended to describe what needs to be done, it is often helpful to provide background information regarding why it needs to be done. This can be handled through an introductory section that addresses the scope of the design, its mission, the types of users to be accommodated and their characteristics, and the types of situations that must be accommodated (e.g., normal, abnormal, commissioning). If this sounds familiar, it should, because we addressed most of these points in Chapter 2, *Designing a New HMI*. We mentioned that, at the end of the analysis phases, one should document findings from those phases in a separate document and get it signed off. If this was done, it is possible to simply refer to this document in the introduction of the specification, et voilà. Otherwise, it will prove useful to capture this information in the specification. One might ask, "Why would we need to capture analysis information in an HMI specification?" The answer is straightforward: the developers may have to make choices or propose changes to the design that was specified because of technical constraints or costs. Understanding the motivations behind the design will help them propose better alternatives or find alternate ways to achieve the same objectives.

Specifying Common Elements: General Section

It is possible to simplify the resulting specification if one takes advantage of a General section. This is similar to factoring elements in an algebraic equation; elements that are common to all, or at least many, terms can often be factored so as to simplify the resulting expression. For HMI specifications, the following elements are suitable candidates for being located in a General section (the reader will recognize many of the following as being part of our earlier definition of HMI architecture; the information collected during the analysis phase of the HMI design process can thus be directly incorporated into the HMI specification):

- Templates used to guide the design process. For example, if we have been able to create templates of status reports, or of P&ID displays, here is a good place to include those.

- Style guide to be used.

- Navigation across HMI components. A good tool to describe this navigation is to use a bubble diagram, where bubbles correspond to individual displays (or groups of displays) and lines between the bubbles correspond to navigation paths. This is sometimes referred to as a navigation map.

- Color, iconic, and other coding used in the various displays. This is also a good place to capture conventions used for displaying indications of stale data, watchdog indication and the like.

- Alarm handling, such as how alarms get acknowledged, what happens when the alarms return to normal, and how an alarm is presented across displays. If required, this section should also discuss any other "exception handling" that may be offered by the application (assuming a software component).

- Buttons with their labels and behaviors. For example, if you have "OK" buttons, describe here what they do and stick to this definition.

- Delays for updating displays and data, as well as for responding to user inputs and requests.

- Physical components such as input devices (e.g., keyboard), pointing device (e.g., trackball, mouse), and display devices (e.g., size, resolution, type). Physical dimensions should also be included, if applicable.

This organization of requirements also helps the specifier ensure that important aspects are not forgotten and keeps the resulting specification compact and easy to understand and maintain.

Specifying Individual Components: Detailed Specification

After we have gone as far as we can with the General section of the HMI specification, we now have to specify individual components. For a software-based HMI, this often means specifying individual windows. A good way to do this is to define and use a format that will be used for each individual window or category of windows. Figure 4–1 and Table 4–1 show a suggested format.

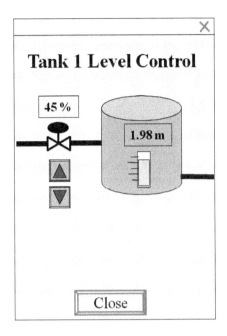

Figure 4–1 *Format for Individual Window or Category of Windows*

Depending on the needs of specification users—for example, the development team—it may or may not be necessary to describe in detail every similar display of a given category. One must always keep in mind that a specification will be a contract that conveys the needs of the organization and users to those responsible for producing the solution; it is therefore important that clarity and rigor prevail. As we mentioned before, a requirement must be written in a way that makes it directly testable; this technique is one of the most important means that can be used to ensure that the specification is clear and coherent.

An Example of Specification

An example of a Table of Contents for an HMI specification is shown in Table 4–2. The reader is cautioned again not to limit himself or herself to this example, as the needs of an individual project may well go beyond what is presented here. Conversely, it is possible that only a subset of what is presented will be required.

Table 4–1 Format for Individual Window or Category of Windows (Reference Figure 4–1)

Requirement number (if any)	If a strict mechanism is used at project level to track requirements and the *downflow*, the requirement number(s) corresponding to the displays should be indicated here.
Display title and number (#)	Each display should have a unique title; adding a display number can also be useful.
Purpose (or associated goal)	Each display should support achieving a goal (e.g., as found in the task model). A display without an associated goal indicates either a useless display or a gap into the supporting analysis.
Description of individual display components	This is where the rubber meets the road. If good communications and a cooperative spirit exist between those specifying and those developing, it will be practical for all involved to stick to a brief, but clear description of elements (those not already covered in the General section) that show on the display. As shown in Figure 4–1, a typical description of display components would comprise: ■ A description of each of the tanks (e.g., labelling, if any, internal level indicator and scale, and numeric indicator, if any). ■ A description of the dials, including the required scale(s) and limits (e.g., alarms). Sometimes, the values on the scale(s) may be specified afterward, but if the information on the range for the variable(s) can be determined at the time that the HMI is designed, then it should be included in the specification. ■ A description of the action of the "up" and "down" push buttons, with limitations, if any. Note that matters such as color, flashing when limits are reached, and label sizes (e.g., for the tank indicators and the dials scales) should normally have been addressed in the General section of the HMI specification.

Table 4–2 Example Table of Contents

Revision History and Versioning
- Introduction
- Scope
- Mission of the system/HMI
- Operational scenarios (based on user's hierarchical task analysis)
- User's characteristics
- Whom to contact to obtain clarifications on this specification

General Aspects
- Style guide
- Navigation map
- Coding conventions
  - Color, iconic and other visual or auditory coding
  - Stale data
  - Watchdog
- Templates
  - P&ID displays
  - Alarm displays
  - Status displays
  - Alarm sheets
  - Individual component control windows
  - Other displays
- Alarm handling
  - Incoming alarm
  - Acknowledging active and return-to-normal alarms
  - Inter-display alarming
  - Filtering
- Buttons
  - Standard buttons
  - Protected buttons
- Delays
- Workstation
  - Input device
  - Pointing device
  - Display device
  - Footprint

Detailed Specification
- P&ID displays
- Alarm displays
- Status displays
- Alarm sheets
- Individual component control windows
- Other displays or componants

Conclusion

References and Additional Readings
The following are references as well as additional sources of information readers may find useful.

Paper-based Publications
IEEE Std 830-1998 Recommended Practice for Software Requirements Specifications, Institute of Electrical and Electronics Engineers. (Probably the best-known and most readable source of guidance on how to write good requirements for software.)

IEEE Std 1233-1988 Guide for Developing System Requirements Specifications, Institute of Electrical and Electronics Engineers. (Quite similar to the previous publication, but with a definite *system* flavor.)

Web-based Resources
http://satc.gsfc.nasa.gov/support/STC_APR97/write/writert.html, some very practical tips on writing good requirements.

5
Improving an Existing HMI

In previous chapters, we have addressed the most difficult situation—namely, starting from scratch to design, evaluate, and specify a brand new HMI. In this chapter, we consider the situation in which we wish to improve, through updates or enhancements, an existing HMI.

Before doing that, however, we should ask the question: "Why do we need to do anything about an existing HMI?" In the next section, we provide answers to this question. Then we propose approaches and strategies on how to improve existing HMIs for two types of situations. One aspect that will not be addressed in this chapter is the systematic improvement of an alarm system as this area is a project by itself. However, the interested reader can refer to the *References and Additional Readings* for suggestions on where to get more information on this specific topic.

Why Improve an Existing HMI?

We may wish to improve an existing HMI for various reasons. In one case, we may be introducing new automation elsewhere in the plant and wish to prevent or minimize inconsistencies between the HMIs that will be introduced and the HMIs of existing systems. This could be achieved by ensuring that the HMIs for the new automation are consistent with those of the existing systems, or perhaps more commonly, by improving the hardware or software, or both, of the HMIs for the existing systems. The methods and strategies provided in the following sections, as well as in the previous chapters, will be useful in doing this.

However, we concentrate here on the case where we wish to continuously update an existing HMI to prevent or close a *results gap* (Figure 5–1). We emphasize this aspect because it is often neglected and usually results in the system not delivering its anticipated benefits over the long term. The results gap is the difference between the current performance of the "operator-HMI-process" system and the near-optimal performance we should normally expect. For example, the initial design may have supported the operator nearly optimally during the start-up of the process, but because of changes in the process, the instrumentation or operating practices, the information provided is no longer relevant. The reasons for this gap may be related to a number of factors. People factors include staff responsible for system performance getting busy on other projects, shifting priorities, or management getting the notion that, once you spend time and resources implementing a system (be it a human-machine system or a purely hardware or software system), it no longer requires upkeep until the end of its useful life. Technology factors include such things as the normal wear and tear of sensors and actuators, with an accompanying drift in performance. Process factors include the normal and expected evolution of operating practices, which may drive a wedge between the support offered to the operator through the HMI and the support required to implement the improved operating practices. Fortunately, it does not have to be like this. By following a simple process, one can not only secure the existing results, but improve the HMI so benefits will keep accruing from continuing improvements. To better understand what we need to do, we distinguish between:

- HMIs for which the design basis exists. These HMIs have normally been designed using either the approach described previously, or another sound approach that nearly optimally matches the design of the HMI to the process management requirements and to the operator's capabilities. They are generally "well-designed;" their design basis exists and is normally documented. Here, "design basis" refers to the underlying knowledge, standards, and rules, as well as assumptions, upon which the HMI design rests. For example, what is the set of tasks for which the HMI has been designed, what are the training assumptions for the users, which coding (e.g., color, symbols) conventions have been used, what kind of validation of the HMI (i.e., usability tests) has been carried out, and which style guide has been used? In particular, if a standard such as ISO 13407 has been used, there are requirements indicating what should be documented. Note

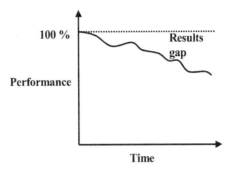

Figure 5-1 *Results Gap*

> that although you may find that the design basis is spread over several individual documents, having this information will simplify and speed up the improvement of the HMI.

- HMIs whose genesis is unknown and for which we cannot locate the design basis. If we are to make any significant improvements to the system, we will need to understand the basis for the existing design and will probably face a bigger challenge than in the previous case.

Because these are two different situations, we need a different strategy to deal with each. In the next section, we will examine these strategies. One word of caution before we proceed: we are talking here about enhancing an HMI, not about redesigning one from scratch. If a complete or substantial upgrade is contemplated, you should treat it as a new design.

Improving an Existing HMI—With Design Basis

Even though it is tempting to believe that an HMI is well-designed, do not assume this is the case if there are no vocal complaints from the operators, as plants are filled with inefficient, error-prone HMIs. Over time, people may have simply accepted them and adapted to the lower level of performance that these HMIs allow them to achieve. If you do not know for a fact that the HMI has been properly designed using an appropriate technique, or if you cannot locate its design basis, you should jump to the next section, *Improving an Existing HMI—Without Design Basis*.

When starting with a well-designed, well-documented HMI, the key point to remember is that the HMI has been designed from the start to support the operator in carrying out his or her tasks. Anything that changes the way those tasks are carried out, or anything that can improve on how the HMI supports them, needs to be identified and dealt with.

It is possible to define a fairly straightforward process to not only prevent the erosion of performance, but also to enhance performance even above the initial value. This can be done by taking advantage of advances in technology or simply by incorporating lessons learned on how to best support the operators through the HMI. Figure 5–2 shows this process and its inputs.

The process above is meant to run continuously, over the life of the human-machine system. In practice, there needs to be a champion, owner, or custodian of the system who will be responsible for ensuring that the system continues to deliver results[1]. Having somebody specifically responsible for continuous improvement is about the only way to ensure that the process will be effective. The main steps of the process are:

- **Data collection**, where the custodian obtains information on changes that may, and likely will, affect the human-machine system. Typical inputs include a brief look at the daily operator log, incident or accident reports (even if informal), new operating procedures (to determine if changes to the HMI are needed to support them) and new or improved operating practices. New equipment may also have to be taken into account if it impacts the operators' tasks or if it offers new data that operators may need to use. Operator surveys may also be quite useful; they take a bit more work to do, but provide other valuable information on what works well and what needs to be improved in the system. It may also be very useful to consider management inputs. For example, the goals of reducing start-up time or improving quality can often be achieved, or partly achieved, by improvements to the functionality and usability of the HMI.

---

1. If one adopts a wide perspective about achieving results, it may be practical to have the responsible individual attend to other, and synergistic, aspects of the human-machine system, such as on-going optimization of the process control application, evaluating the performance of the control loops, etc.

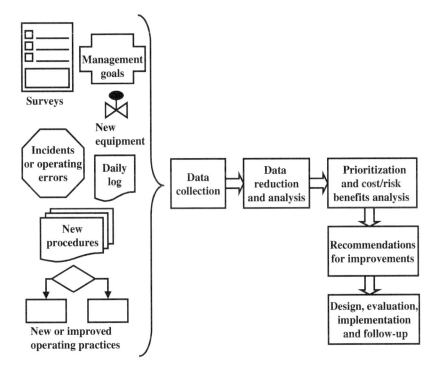

Figure 5–2 *Process to Enhance an Existing HMI—with Design Basis*

Obviously, knowing what kind of data to collect will have a big impact on the outcome, so some basic training in human factors engineering, or even in usability engineering, would be helpful here.

- **Data reduction and analysis,** where data collected previously is examined to determine if there are genuine issues, or opportunities to improve benefits, through changes to the HMI. Changes should not be made because a user, or even some users, say a change should be made. A thorough examination must be carried out to ensure that the requested change will improve operations, or safety, or both. Also, care should be taken to avoid turning this step, as well as the previous one, into a *paper exercise*, meaning that the process should not be driven solely by written records, but also by an active dialog with the various types of HMI users. Depending on the urgency of the situation, candidate findings should then be summarized either periodically (e.g., monthly) or on an ad hoc basis.

- **Prioritization and cost-risk-benefits analysis**, where findings from the previous steps are scrutinized to determine if a business case can be assembled for making changes to the HMI. By *business case*, we do not advocate a lengthy exercise. Simply put, we need to answer the question: "Is it worthwhile to take action, based on the data that has been collected?" For example, a once-every-two-years task of limited scope and complexity may not warrant a change to the HMI. On the other hand, providing a few additional pieces of information on the main display used for start-up may cost $1,000, but this investment may be recouped within the first month of operation. Another concern is, of course, safety—for the environment (including the neighboring population), staff and equipment. If a change to the HMI reduces the risk of occurrence of an anomaly, there may be a fairly simple business case that can be made to justify it. In one instance experienced by the author, a small change (a simple annunciation window) was implemented onto an HMI to prompt the operator to carry out a frequent and seemingly trivial task on a paper machine. This small change had a payback of a few hours, yet it provided an additional net yield for this paper machine that amounted to the equivalent of almost two full days of production each month.

- **Recommendations for improvements**, where recommendations, which appear justifiable, based on the previous steps, are presented to management for approval.

- **Design, evaluation, implementation and follow-up**, where the approved changes are put into application. At this point, it may be tempting to skip some of the steps of the HMI design process, especially if the initial design using the method outlined earlier has been very successful. Doing this would be a mistake, for several reasons. First, if the method has been successful before, it is likely to be successful again. Second, most changes pose a certain amount of risk. For example, removing pieces of information from displays, adding or deleting displays, changing the alarm management philosophy, etc., could easily lead to unexpected situations. Usability testing provides a certain assurance that changes will not create new, unexpected, difficulties. Another reason is somewhat more subtle. As we have seen, an HMI is an integrated set of functionalities aimed at providing the required information and controls to a user. Changes to the HMI must be kept coherent with the overall HMI architecture. It is dangerous to introduce exceptions or inconsistencies into the HMI. The risk of introducing these exceptions and inconsistencies is minimized if the HMI design method is used. Obviously, the effort to design, evaluate and specify changes to the HMI will likely be substantially less than the effort required for the initial design.

This process may appear comprehensive, and it is. This does not make it time-consuming, however. It should run continuously. It should be expected to require no more than a few hours each month for the data collection, reduction, and analysis steps. Only when significant changes appear warranted should effort be expended to provide estimates and to draft work plans.

Improving an Existing HMI—Without Design Basis

Contrary to the previous case, the design basis for an existing HMI may be unknown or undocumented. This is often the case when the system was procured from a supplier as a "turnkey" system or if one has inherited a legacy system. Even though the HMI design may reflect current practices and may meet a number of operational needs, there is usually ample room for improvement. The issue can be tackled in a number of ways, but for best results, the process shown in Figure 5–3 is recommended.

As is apparent from Figure 5–3, the bottom part of the process is quite similar to the one used for improving an HMI for which we already know the design basis. The main difference here is that, ideally, we should reconstruct this design basis, at least to some extent, as shown in the top part of the figure. This means, at minimum, to clarify the organization's expectations regarding the mission of the HMI and who its intended users are. This will prove useful if we need to modify the HMI later. For example, if we are to add or modify information or functionality, do we need to cater to the needs of I&C technicians or process control engineers as well as to those of the operators?

One should also try to sketch the overall HMI architecture. This is often most easily done by drawing a navigation map of the HMI.

90   Chapter 5—Improving an Existing HMI

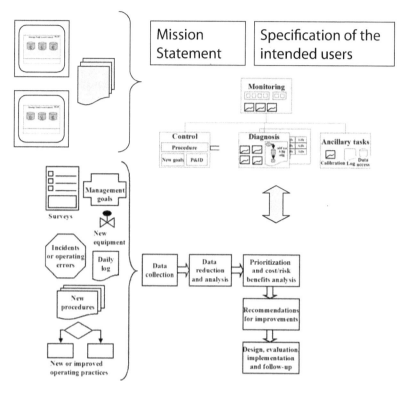

Figure 5–3   *Process to Enhance an Existing HMI—without Design Basis*

Because there can be a large number of displays, it is often possible to obtain a useful map that will show the navigation between categories of displays rather than between individual displays. We already mentioned in a previous section that coherence is an important feature of good HMIs. Going through this exercise will help provide an idea of the size of the HMI in terms of displays (something that is rarely known beforehand) and will also help in getting familiar with the overall look and feel of the HMI. While documenting the HMI architecture, it is useful to pay close attention to functionality that may exist in the system but has fallen into disuse because there was no training or because staff has simply forgotten about it. Legacy systems often embody a number of good ideas and it is not uncommon to find useful features that are not being taken advantage of[2].

---

2. This is not unlike the situation of modern software packages where, typically, only a small portion of the available features is used.

One should be opportunistic in obtaining information to carry out the activities outlined above. Typical sources include existing operator or training documentation, information that can be gathered from the system's supplier, and discussions with the parties involved in the design and procurement of the system. Trainers are also a great source of information. The existing HMI is also often a good source of information. For example, printing a representative sample of the displays and examining various menus and sub-menus often provides enough information to draw a reasonable first version of the HMI architecture.

It is difficult to estimate the effort required to understand and to document the design basis of an existing HMI, but it is safe to assume that, for most cases, this should not take more than a few days of effort. For example, in the author's experience it is often possible to document the design basis of an existing HMI of up to 50 to 100 displays or panels in a week, give or take. Doing the same for a much larger system, say ranging from 500 to 5000 displays or panels does not take much longer as there are often not many more categories of pages. Rather, there tend to be more pages in each category and the design basis is usually done for a category of pages, rather than for an individual page (except for cases where a given page was designed in response to a specific need).

Once the design basis of the HMI has been documented, it is possible to use the second part of the process, which is the same as was described in the previous section. A few comments are in order at this point:

- There will likely be a greater number of enhancements that will be identified for an HMI that has not been designed from the start by the method described previously. In particular, a number of inconsistencies are likely to be found in the design. All candidates for enhancements should be identified, but care must be taken to follow the process from the previous section, in particular those steps having to do with prioritization and cost-risk-benefits analysis. It will probably not be possible to correct everything, so it is important to concentrate on those items that will enhance productivity and safety.

- Documenting the HMI architecture will not only provide a sound basis for enhancing the system, it will also supply valuable information for specifying the next revision of the system or even its replacement.

- A typical outcome of the reconstruction of the design basis of the HMI is identifying training needs that should be addressed to maximize the results that can be achieved through the existing HMI. Further, the exercise will also provide tools that can be useful in training (e.g., the navigation map).

As can be seen from the above, improving an existing system is technically and administratively straightforward. It is often a good practice to avoid looking at this activity as a maintenance job where resources must be allocated to keep up the system's performance. Instead, try to look at it as an area for continuous improvement. As hard as we may have tried to design the best human-machine system possible, long-term operating experience will provide useful information on new and better ways to achieve the organization's productivity and safety goals. The processes explained here should be used to incorporate this experience into the human-machine system.

Lastly, it is important to realize that changes or enhancements to an HMI will have consequences for the users. Changes thus need to be adequately conveyed to the users and considered for initial and continuous training.

Transition

In this chapter, we have defined two approaches to continuously improve HMIs, so as to ensure that the "human-machine" system continues to operate nearly optimally.

Up to now, we have considered the HMI design problem as being the same, irrespective of the type of situation. While this is generally true, there are some peculiarities associated with the type of process for which we design the HMI (e.g., continuous, batch, hybrid). This will be the subject of the next chapter.

References and Additional Readings

The references for this chapter are the same as for the previous chapters as we are simply re-applying existing rules and principles. However, the reader may be interested in the following paper:

Salo, L., Laarni, J. & Savioja, P. *Operator experiences on working in screen-based control rooms*. The 5th ANS International Topical Meeting on Nuclear Plant Instrumentation, Controls, and Human Machine Interface Technology, Albuquerque, NM Nov. 12–16, 2006. (Even though this paper deals with nuclear applications, it provides useful feedback on the "human" effects of modernization and upgrade projects. In particular, it discusses the fact that even technically successful projects may present new human challenges.)

Also, readers interested in alarm system improvements should refer to *EEMUA 191 Alarm Systems—A Guide to Design, Management and Procurement*, produced by the Engineering Equipment & Materials Users' Association. (This guide contains a section on managing an alarm improvement program, and a selection on buying a new alarm system.)

6

Continuous, Batch, Discrete and Hybrid Applications

Modern monitoring and control systems are used in a variety of applications; here, *applications* means the monitoring and control of continuous, batch, discrete, and hybrid (combining the previous types) processes. Because of this variety, there is often a conception that HMI designs will vary, depending on the type of application. This is both true and false, as we will discuss in the next sections. We will also provide additional design guidance and identify potential pitfalls.

Designing HMIs for Different Types of Applications

HMIs—whether for continuous, batch, discrete or hybrid applications—often require some specialized displays and control functionalities. Further, because of intrinsic differences in the processes, there may be additional considerations that could impact HMI design, such as:

- The information required to monitor and control continuous processes is necessary to ensure that the process runs smoothly and within quality parameters. After a successful start-up, the operator should primarily aim to monitor and control the process in a steady state. Major process changes (e.g., changing the key characteristics of what is being produced) occur relatively infrequently; small changes to keep the process within bounds occur more frequently.

Monitoring the achievement of goals occurs at the overall process level.

- The information required for batch processes is needed to ascertain that each batch achieves the prescribed quality parameters in the allocated period of time. Start-up and shutdown of a batch process occur every time a new batch is produced. Changes in required product characteristics occur mostly between batches, rather than within a batch. The monitoring of the achievement of goals occurs mostly at the individual batch level.

- Discrete processes such as manufacturing can often be seen as a continuous flow of small batches (e.g., small sub-assemblies flowing from one cell to another). Specialized devices such as artificial vision devices, robotic effectors, etc. may be used, potentially resulting in the need for specialized visual representations. There is also frequently a challenge with tracking a large number of products in various stages of transformation. Depending upon the value or importance of the individual units, it may even be necessary to track the status of each unit. Monitoring often occurs at the unit level as well as at the overall production level.

- Even though there are differences in the relative percentages of the various types of automation (e.g., programmable loop controller versus proportional-integral-derivative controller) used in the different situations, the same types of devices are generally used. Exceptions do occur, and may take the form of the vision, robotics or other (e.g., conveyors) devices mentioned previously. However, they do not fundamentally change the HMI's information content.

The impact of the above differences often results in variations in the organization of the information and in navigating among the different pieces of information available to the operator, rather than in the need for radically different types of information. It is also worth noting that the process used for designing the HMI is the same, irrespective of the type of application.

It is now possible to derive design principles that will yield a better HMI design for each type of application discussed above:

- For continuous processes, a key principle is to organize the information to provide support for monitoring and controlling the process as a whole. Other key organizing principles are to capitalize on the continuous flow of mass, energy and, in some cases, information, all with respect to time.

- For continuous processes again, one should pay attention to the fact that when the process is not running, the sub-systems of which it is composed are usually *loosely coupled*, and may even be *decoupled* from one another. Coupling refers here to the exchange of mass, energy or information between the sub-systems. When the process is running, though, its sub-systems are normally strongly coupled. It follows that the HMI must support the need to couple or synchronize the sub-systems that make up the process during start-up and other important transitions. This differs from the frequent use of sub-systems as the natural border for the content of displays, and may require displays that help the operator deal with several systems simultaneously.

- For batch processes, the information should normally be organized to help the operator efficiently set up a batch, start the process, track its progress, and then shut it down and prepare for the next batch. Overall process monitoring can then be accessed as needed; easy access to recipe information and associated controls is often important.

- Discrete manufacturing processes combine features from both continuous and batch processes. Choosing how to organize the information will depend on the relative importance of the unit being worked on and of the overall process. For example, manufacturing a large volume of low-value units will likely lead to greater emphasis being put on overall process monitoring, while producing high-value units (e.g., an engine rebuilt process) will likely increase the value of monitoring individual components. If several similar production lines or cells are present, it may also be especially worthwhile to capitalize on visualization techniques, such as object displays, to provide the operator with easy means to overview several process variables at once (see Appendix). Lastly, special attention must also frequently be given to emergency features (e.g., stops) that may be mandated by local or other regulations.

All of the design principles and rules provided in Chapter 2 regarding the HMI architecture still apply and should be used along with the organizing principles just provided.

As can be inferred from the previous discussion, those organizing principles will mostly affect the architecture of the HMI and navigation among its main groups of information. Practically speaking, information required to support the monitoring and control of process- or batch-level goals will be identified, as will the means to navigate between these and other groups of information. Figure 6–1 shows where those principles are taken into account in the HMI design process. There will be minimal impact, if any, on the design of individual display elements (e.g., trend graphs, digital displays, alarm messages).

Transition

In this chapter, we provided additional design guidance for various types of applications. A further challenge that the HMI designer will face is how to integrate different types of HMI components; this will be addressed in the next chapter.

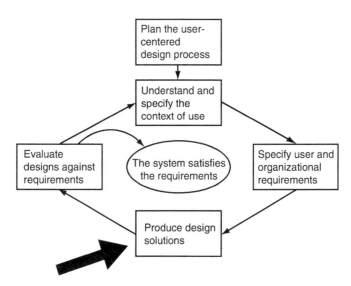

Figure 6–1 *Impacts of Application-specific Design Principles on the HMI Design Process*

7

Integrating Heterogeneous HMIs

The HMIs that we design are typically implemented using an aggregation of computerized and hardwired panels, mobile devices, operating documentation, and large screen displays. Additionally, the designer frequently has to contend with a variety of platforms. *Platforms* refers here to either hardware (PCs, DCSs, PLCs and even wireless devices) or software (Windows, Java, Linux, Web, or other specialized software tools). Effectively integrating these components to best support the operators requires that the designer pay close attention to a number of factors. In this chapter, we will first propose a definition of *integration*. We will then discuss integration principles and rules for common types of HMI components.

What Integration Means

Before diving into this topic, we will define integration as "the combining of heterogeneous HMI components in such a way that human error is reduced, performance is optimized, and training requirements are minimized."

To get a better picture of the challenges facing us, consider Figure 7–1, showing a situation where process monitoring and control is spread across several levels.

The HMI designer has to contend with a situation where several classes of personnel will use a variety of HMI components to operate the plant as economically and as safely as possible. The following

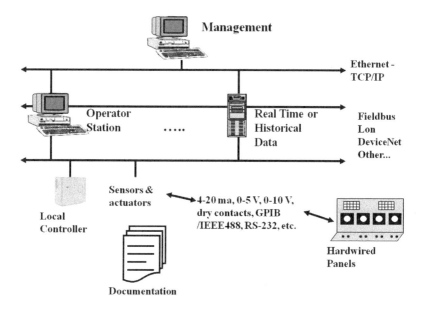

Figure 7-1 *Example of a Heterogeneous Environment*

sections deal with several aspects related to integrating these heterogeneous components.

Simple Integration Strategies

There are some very simple, inexpensive, yet effective strategies that should be used early in the project, usually in the planning phase. Applying these strategies will help operators and supervisors ensure that they monitor and control the right things. The strategies consist of:

- Defining and strictly enforcing a common terminology across the various HMIs, whether they are computerized or hardwired. You must also ensure that the documentation (including the procedures) uses this terminology. A good way to achieve this is to develop and maintain a dictionary of terms that are permissible and to pass it on to suppliers as part of contractual requirements. You should also make the dictionary available to the training staff for incorporating the terms into the training.

- Adopting a color-coding convention that will be uniform among the various components of the environment. This color scheme should encompass the coding of component states (e.g., opened, closed, run, stop), process components (e.g., liquids, gas), alarms, annunciation messages, etc. It should be duplicated (to the greatest extent practical) on the electronic and hardwired panels. When using color, to always supplement color with another type of coding to cater to the needs of color-impaired staff.

- Consistent units. Several incidents and accidents have occurred because different units (e.g., Imperial, International System) were used in different parts of an HMI. Consistency and clarity should be strictly enforced in this aspect of the design.

- Adopting conventions regarding font styles and sizes. Alternatively, define reading distances for various components of the HMI and specify font sizes and types that designers should use.

- Ensuring that the transfer between automatic (computerized) and manual (hardwired) devices provides non-ambiguous and adequate feedback to the operator, to avoid any misinterpretation as to how the process is being controlled.

- Ensuring that alarm and annunciation acknowledgement are consistent between the computerized and hardwired panels. The number and location of the point(s) of acknowledgement should also be selected such that they best support what the operator needs to be informed and to efficiently acknowledge the alarms.

- If different computer-based systems are used, ensuring that the style guide upon which their HMI is based, to include the grouping of the information, navigating through it, and all of the coding conventions, is as consistent as possible.

The above are simple, low-level, and inexpensive integration strategies. They are not sufficient by themselves, but are representative of simple things that can be done to achieve a better design.

Component-specific Integration Rules

The strategies provided in the previous section provide generic and broadly applicable guidance to achieve a degree of consistency in a composite HMI. There are additional considerations that should be taken into account relative to the characteristics of components of the HMI. In particular, the following questions need to be addressed:

- Which control or monitoring functionalities should be allocated to either the computerized or the hardwired panels?

- When should a large screen display be used?

- There is a growing availability of wireless or mobile devices in the market; how can they best be used to support the operator?

These are complex questions, and current state-of-the-art human factors engineering knowledge does not always provide satisfactory answers. However, some guidelines can be provided that will help in designing better HMIs. These are discussed in the next sections.

Allocating Control and Monitoring Functionality to Computerized or Hardwired Panels

The decision to allocate control, monitoring, or both types of functionality to either computerized or hardwired panels is often made in the process industry because of cost considerations. Interestingly, other industries, such as aerospace or nuclear, are still somewhat reluctant to use what they call *soft controls*[1].

There are a few guiding considerations and principles that should be used in deciding on the allocation:

- **Safety and reliability issues.** First and foremost, safety and reliability should be considered in deciding whether a

---

1. Two of the reasons for this seem to be the need to gain regulatory approval prior to making any significant design changes which may force the vendor to go through long certification processes, and the relative lack of operating experience with those types of controls in those high reliability industries.

function should be performed through computerized or hardwired panels. There are some very reliable computerized systems, but the reliability of a properly designed hardwired panel may make it the tool of choice for safety-critical controls such as emergency shutdown. Also, speed and ease of access may lead to allocating some functionality to hardwired panels.

- **Ability to maintain an awareness of the state of the plant.** In the not-so-old days, operators would routinely walk the panels when carrying out a specific control or monitoring task. While so doing, they would often perform *monitoring by wandering*, casually looking at indications on the panels as they walked by. This helped them maintain a general understanding of the status of the plant. This activity was helped by the generally parallel presentation of information, made available by virtue of the fact that the sources of information were present side-by-side on a hardwired panel. The ability to casually monitor the plant has been somewhat lost with computerized panels, which typically provide operators with a more serial view of the plant. To successfully integrate computerized and hardwired panels, it is paramount to ensure that the operator can still—and with an absolute minimum requirement for navigation or effort in retrieving information—maintain an awareness of the general status of the plant.

- **Perceived complexity.** With the increase in the span of control and monitoring responsibilities assigned to the operator, there has been growth in the number of indicators and actuators that must be attended to at hardwired panels. This growth has often resulted in a perceived increase in the complexity of the panels. Computerizing is sometimes seen as a way to reduce this apparent complexity. This is sometimes successful, but it is important to realize that reducing the number of discrete indicators and controls may sometimes just displace the problem; it is possible to go from a bad situation (too many indicators and controls) to a worse one (too many indicators and controls, grouped onto too many (or hard-to-reach) computerized display pages). There are a few suggestions that can be made to avoid this. For example, before computerizing for the sake of reducing the apparent or perceived complexity of a hardwired panel, take a step back and ask yourself whether you will reduce the workload of the operator, displace it, or actually increase the number of steps that he or she will have to carry out to monitor or control the process. Another question to consider is whether you will render part of the process more obscure. If this is so, but for some reason you still want to provide a computerized panel, you need to design it in such a way as to allow the operator to remain aware of the process; trying to achieve this by simply adding one more alarm to the list of system-generated alarms will not generally do. Lastly, consider the possibility of supporting a whole task, rather than moving part of the functionality needed for carrying out a task. For example, if you choose to provide a computerized panel to monitor and control part of a system, consider designing a panel that allows start up, shut down, changing setpoints, etc., for the system rather than forcing the operator to use both the computerized and hardwired panel to perform those tasks.

- **Turning data into information.** For the purpose of our discussion, we will consider information to be the set of data required to carry out a given task. Hardwired panels tend to be rich in data but poor in information. This is because of the one measure (one display philosophy) that has traditionally been used for designing these panels, and also because the layout of the data does not clearly match the tasks to be carried out by the operator. One advantage of computerized panels is their ability to easily turn data into information. Consider, for example, determining whether a fluid is in liquid or vapor phase. This can be achieved by simultaneously considering the temperature and pressure of the fluid and combining this information with its saturation curve. Many engineering students have spent stressful moments during exams trying to do just that, sometimes with disappointing results. The job of the operator would be no less complex. Doing that with a computerized panel can be made very simple, as shown in Figure 7-2. The task of determining whether the fluid is in vapor or liquid phase has been moved from an analysis task, using individual measures and a vapor table, to a perceptual one by virtue of the a priori integration of data to produce information that supports the operator.

The question is then how to turn data into information. This might be a difficult question, but if we look at the problem from the operator point of view, what differentiates data from information is the use that he or she makes of it. That is, if the operator needs to make a

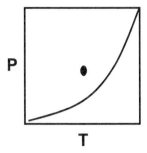

Figure 7-2 *Information Integration*

decision on a rate of change, then a series of instantaneous values is data, whereas the actual calculated rate based on those values is information. Similarly, if the operator needs to know whether all of the permissives on a piece of equipment are in their ready state, the individual permissives constitute the data, but the status of the piece of the equipment is the information sought by the operator. The tool to identify what is the information is thus the hierarchical task model that is normally produced in the course of the HMI design.

Using Large Screen Displays

Whether to use large screen displays (LSDs) as part of one of the HMIs is still a hotly-debated matter in several places. There are diverging views about the usefulness of an LSD. Once a decision has been made to use an LSD, the next question often becomes, "What will we use it for?" This is a difficult question, especially since previous studies have shown that how a particular media is used may evolve substantially over time. Even though it has been said (with sometimes a surprising amount of truth!) that LSDs are targeted more at visitors than at the operating staff, given the right method it is possible to design LSDs that are useful. Also, LSDs enable important control room visitors (e.g., control technicians or process engineers) to watch for process status, without getting in the way of process operators.

There is also an interaction between the type of technology used in the LSD and how the LSD is used. Table 7-1 reviews the pros and cons of various technologies.

Table 7-1 Comparison of LSD Technologies

Technology	Pros	Cons
Front projection	<ul><li>Flexibility in modifying the content.</li><li>Compact.</li><li>Relatively inexpensive for larger displays.</li></ul>	<ul><li>If a person steps in front of the projector, the LSD loses its usefulness.</li><li>May be susceptible to vibration.</li><li>Legibility may be affected by strong ambient lighting.</li></ul>
Rear projection	<ul><li>Flexibility in modifying the content.</li><li>Performs well even if staff stands in front of LSD.</li><li>Relatively inexpensive for larger displays.</li></ul>	<ul><li>Requires a fair amount of space behind the LSD for the projection apparatus.</li><li>May be susceptible to vibration.</li><li>Legibility may be affected by strong ambient lighting.</li></ul>
Hardwired (this, in effect, is a special case of a hardwired panel)	<ul><li>Robust.</li><li>Compact, as it does not require either back or front clearance for projection apparatus.</li><li>Initial cost can be low if LSD is simple.</li></ul>	<ul><li>Lack of flexibility, thus difficult to keep up to date or in presenting context-sensitive content.</li><li>Cost may escalate quickly for large LSDs, frequent updates or content-rich LSDs.</li></ul>
Solid-state or liquid crystal based	<ul><li>Flexibility in modifying the content.</li><li>Robust.</li><li>Compact, as it does not require either back or front clearance for projection apparatus.</li></ul>	<ul><li>Depending on technology used, legibility may be affected by strong ambient lighting.</li><li>Cost may be high for larger units.</li></ul>

The following guidelines will help determine how to use LSDs as part of a set of HMIs:

- LSDs are especially useful when information must be shared among various individuals; for example, while diagnosing a problem.

- LSDs provide a useful means to access information if an operator must move around in the control room yet maintain an awareness of specific information.

- There is often the conception that because an LSD is large, it is possible to display a large amount of material on it. This is not entirely true, since LSDs are normally read from a distance, thus information must be presented in a fairly large format to be legible. The overall amount of information that can be shown on an LSD may not be significantly more than what can be shown on a regular display.

- A potential, and important, role for an LSD can be to provide information to the operating staff regarding achieving operating and safety goals for the process.

Detailed design rules for LSDs can be found in the Appendix.

Using Portable and Wireless Devices

Early forms of wireless devices have long been used in process plants (e.g., walkie talkies). However, modern devices provide the ability to display sophisticated information in text and graphic formats. More and more, portable and wireless platforms such as pen computers, tablets, personal digital assistants, etc., are used to provide information and sometimes control functionalities to a roving operator. Specific requirements and limitations (e.g., bandwidth, display real estate) often exist for these platforms. Proven principles and rules for best practice as part of a total HMI solution are scant, but a few generic guidelines can be formulated. Here are some recommendations on using wireless HMIs and wireless technology:

- Device-specific style guides often exist and are available from manufacturers for wireless and portable devices. They provide valuable guidance on the best design practices to be followed for these types of devices. For example, guidelines on display density or menu structures often vary between PC-based displays and mobile displays because of restricted communication bandwidth and latency. Typically, very dense display designs are preferred for wireless applications so as to minimize the number of data queries that the user must carry out. Further, device specific style guides will usually address the preferred key assignment, means of pointing, data entry, etc.

- Wireless devices should not be used for critical operations, as the technology is still generally susceptible to electromagnetic noise.

- The legibility of the display should be assessed in the actual context of use as wide variations exist between different screen technologies.

- Designing and validating a mock-up of the HMI is going to be even more critical for wireless devices as they tend to have less-powerful interface resources. For example, screen real estate, computational resources and memory are often severely limited; navigation is also notoriously more difficult and laborious.

- Wireless devices should not be used as the main or sole means to access plant information, in most situations.

- Wireless devices can be used to provide point information and to reduce the amount of walking that operators have to do.

- Wireless devices can be used to provide access to electronic documents and to capture field information, thus potentially eliminating the need to write the information down and later enter it into a computer. This may reduce human error and increase personnel efficiency.

Applying the HMI design process, in addition to using the right style guide, will pay off handsomely, as it is very difficult to strike the right balance of functionality and usability in a wireless device.

Combining HMI Resources

Given the variety of means discussed so far, it might look like a daunting task to orchestrate the different media into a harmonious whole. This does not have to be so.

A key to successfully blending the various ingredients of a good design is to leverage the notion of HMI architecture introduced in Chapter 2, which discussed designing new HMIs. Remember that thinking in terms of HMI architecture allows you to have an abstract representation of the needs of the operator in terms of monitoring and control functionality, independent of how the HMI will be implemented. Once you have formulated a clear picture of monitoring and control functionality needs, preferably in hierarchical form, and have a better understanding of other requirements and constraints you may be facing (e.g., safety, redundancy), it becomes easier to choose which HMI components (e.g., computerized, hard-wired, LSD, wireless) are best allocated to them.

Our last word of advice is to rely on the HMI design process that was outlined earlier. Once you decide on a certain combination of HMI components, to which you have allocated certain roles, you should design a mock-up, test it and then improve upon it. Keep in mind that you, or anybody else, are not likely to get the optimal solution on the first try.

Transition

In this chapter, we proposed integration principles and rules for various types of HMI components. The reader should, however, bear in mind that integration is a complex issue for which state-of-the art knowledge is still imperfect; further principles and rules are likely to emerge. The next step is to discuss how those HMI components will be organized in the control room, or more appropriately, the control center. This will be addressed in the next chapter.

References and Additional Readings

The following are references as well as additional sources of information readers may find useful.

Paper-based Publications

Several of the references mentioned in the previous chapters (e.g., Nureg 700, Mil Std 1472F) are relevant to LSDs.

Web-based Resources

Regarding other technologies (e.g., mobile user interfaces), the style guides and other design guidance tends to be either manufacturer specific or to depend on the type of operating system that is used. Nevertheless, the reader may find the following links useful.

http://www.stcsig.org/usability/topics/wireless.html, the site of a comunity on usability and user experience from the Society for Technical Communications.

http://www.w3.org/Mobile/, the site of the Mobile Web Initiative, from the World Wide Web Consortium. The Initiative's aim is to ease and support mobile access to Web content.

8

Overall Organization of HMIs in a Control Room

So far, HMIs have mostly been considered in relation to a single user except, perhaps, for our discussion of large screen displays (LSDs). This perspective needs to be broadened for two reasons:

- HMIs are normally embedded into workstations that are frequently used within *control rooms*.

- Practically speaking, even though a single operator will often carry out independent tasks, team performance is of great importance, and HMI design needs to take it into account.

Further, in a realistic work setting, HMIs used in a plant would also include, by definition, all interfaces between operators and the information required for monitoring and controlling the plant. The following are examples of those HMI components, beyond the more usual computer-based or hardwired panels, LSDs or even portable or wireless devices:

- Operating manuals and procedures.
- Public announcement (PA) systems.
- Phone systems, both internal and external to the plant.
- Printers used for shift reports or alarm printouts.

- Information about completed or in-progress jobs (e.g., maintenance jobs, past or in-progress process upsets, operation and maintenance logs).

- Any ancillary functionality required (e.g., breaker panels, diesel generator control panels).

To provide guidance on how to best organize the HMI components in a control room, we will first discuss the characteristics and governing principles of the use of different types of workstations as those frequently contain the computer-based or hardwired panels that have been discussed in previous chapters. We will then discuss organizing principles for other HMI components. Lastly, we will describe a simple process that can be used to locate the various HMI components in the control room so as to optimize the overall performance of the human-machine system. Note, however, that we will refrain from discussing the design of control rooms as this is an entirely different topic.

Workstations

Workstations as discussed here refer to either seated or stand-up workstations commonly used to host hardwired panels or computerized interfaces that make up the various HMIs[1]. Each of the seated or stand-up workstations (also referred to as consoles) have advantages and disadvantages, as well as peculiarities in application. There are also general principles that apply. These are all discussed in the following sections.

Advantages, Disadvantages and Utilization

Table 8–1 provides general descriptions of the characteristics of those workstations.

Each type of workstation is seldom used alone. Typical arrangements combine seated workstations for long-term monitoring and control, with stand-up workstations to provide complimentary or backup functionality.

---

1. Sit-stand consoles exist that can be used for long-term (several hours) monitoring and control tasks but these consoles are rarely used in a control room setting and will not be discussed here.

Table 8-1 General Characteristics of Workstations

	Type of Workstation	
	Seated	**Stand-Up**
Illustration(s)	Wrap-around workstation Vertically stacked workstation	Status and Annunciation Displays Controls Stand-up workstation
Best Used for	▪ Long-term (several consecutive hours) monitoring and control. ▪ Well-suited to computer-based HMIs but also suitable for hardwired panels.	▪ Mostly used for hardwired panels but may also house computer-based displays.
Other Considerations	▪ Need to provide reading surface for documents. ▪ Even though Mil Standard 1472 F proposes standardized dimensions for seated consoles, this has become less popular, as several providers now offer easily customizable products that are fit for this type of use. If those products are used, it will be particularly important to specify proper dimensions (e.g., height, depth, width, leg clearances), rather than to rely on the product's "out-of-the-box" characteristics.	▪ Mil Standard 1472 F proposes standardized dimensions for stand-up consoles. ▪ The reserved areas (status and annunciation, displays and controls) help ensure best visibility and a lower likelihood of accidental activation.

Organizing Principles—Workstations

While the topic of detailed workstation design is outside the scope of this book, we will review several principles that can be used to choose and organize the appropriate workstation:

- If a decision has been made that operators will spend most of their time in the control room, seated consoles should be considered. In this case:

    - Use the 20–80 rule to arrange components at the workstation, i.e., locate the 20% of HMI components (including procedures, phone system, PA) that the operator will use to carry out 80% of his tasks within easy reach. Easy reach means that the operator should be able to access those components without having to stand up or to extend his arm completely (e.g., to get a heavy binder of procedures).

    - Ensure that you have plenty of space for the phone system, as this is often the most commonly used channel of information transfer in a control room. By the same token, ensure that you have sufficient phone lines available to the operator to handle unexpected situations. Getting a "busy" signal during a process upset can be a very distressing situation.

    - There are detailed design requirements regarding the ergonomics (e.g., height, depth, width, clearances) of the consoles; those requirements should be used (see the section on References and Additional Readings at the end of this chapter). At the same time, beware that blindly following standards and guidelines without taking into account the context of use often leads to poor design. To guard against this, and once you have produced your first design, make a mock-up of the projected workstation or console and try it out. Mock-ups can be very simple; for a sit-down computerized workstation, you can make a good mock-up using a table, a chair, a keyboard, a mouse or another pointing device, and cardboard components to account for the dimensions of the monitor or other components. Such a mock-up is usually sufficient to assess whether the new design suits the user's needs.

- In some legal jurisdictions, there may be regulations (e.g., regarding ergonomics requirements) that must be considered.

- Ensure that the orientation of the workstation with respect to the windows does not induce glare issues. To do so, avoid locating a window behind the operator. If a window is present, make sure a blind is available if glare becomes an issue.

- Similarly, avoid putting a seated workstation right in front of a window, as differences in brightness between the HMIs and the outside light may induce visual fatigue. Again, a blind often provides an easy and inexpensive solution if you do have to locate the workstation in front of a window.

- You should consider furniture made for continuous industrial grade applications. Regular office furniture (e.g., desk, chair) tends to wear out early during the day in-day out use typical of industrial applications.

- If operators are expected to be mostly mobile and spend little time in a control room, then stand-up consoles may offer space saving, thus leading to more-compact control rooms. For stand-up consoles:

  - Locate the status and annunciation, displays and controls panels in the corresponding zones as shown in Table 8–1.

- If seated and stand-up consoles are used in the same control room, then:

  - The rules already mentioned for seated workstations generally apply.

  - Ensure that frequently used controls are available at the location where they are expected to be most frequently used.

  - Ensure that information displayed on the stand-up consoles (e.g., process variables, annunciation tiles) is legible from the seated consoles if the operator is expected to see and understand this information. Remember that legibility can be analytically derived from character height, as described in our discussion on visual perception.

Several other general principles apply regardless of the types of workstations that are used, such as:

- Ensure that maintenance as well as instrumentation and control (I&C) staff can do their work without getting between the operator's most frequently manned workstation and stand-up workstations. For example, ensure that stand-up workstations can be accessed from behind for maintenance purposes.

- If a hardwired panel is used to provide emergency, redundant, or shutdown capabilities, it should be positioned close to the location where it will be used.

- Provide sufficient clearance between stand-up workstations and hardwired panels that are facing one another; here again, detailed guidance is provided in applicable standards.

- It may be useful to delineate no-crossing areas for non-control-room staff near the panels to avoid accidentally blocking line-of-sight or inadvertently actuating controls. Several means have traditionally been used to achieve this, such as a physical barrier or differently colored carpet or tiles in front of, and around, workstations. A combination of these means, as well as awareness by plant staff, probably works best.

- Ensure that there is room for some expansion. Over time you may find you need additional space for documents or for an additional monitor. The current rule of thumb is to allow at least 25% room for expansion with respect to the current footprint of the workstation (including the area where the operator sits). You may want to be more conservative and allow for more.

Additional matters such as how to best organize and locate operators' and supervisors' workstations (e.g., side-by-side, L-shaped), are covered in considerable detail in existing documents (see the Section References and Additional Readings). However, pay close attention to the issue of inter-operator cooperation. This topic is frequently neglected in choosing a suitable arrangement for the consoles. For example, having a linear, side-by-side arrangement as shown in Figure 8–1a may not always yield a good solution when two operators who spend most of their time working independently have to closely cooperate and share the same information during certain operational phases.

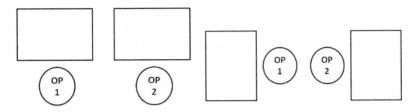

Figure 8–1 *(a) Side-by-Side Arrangement. (b) Alternate Arrangement.*

Alternative and innovative arrangements may work best; for example, having the consoles facing one another, as in Figure 8–1b, rather than side by side allows each operator to remain aware of the information on both consoles merely by turning his or her head. Here again, producing a mock-up and conducting usability tests will pay off handsomely, as it is much cheaper and faster to correct the mock-up than to modify the final design.

Organizing Principles—Additional HMI Components

In addition to workstations, other HMI components must similarly be organized so as to optimize the performance of the operating staff into control rooms and control centers. Fortunately, organizing principles also exist for several of those components:

- If the choice has been made to add an LSD to the control room, then it needs to be located in such a way as to best support its intended role. Some suggestions are:

  - The starting point for locating the LSD, in much the same way as any other equipment, is to have a clear idea of its purpose and then to choose a location that allows the intended users to achieve their goals.

  - The content of the LSD should normally not be a duplicate of the content of the computer-based or hardwired HMIs. Rather, it should compliment them. One often hears that an LSD best provides an overview of plant condition or status. This can be achieved by tailoring its content to reflect the degree of achievement of operating (efficiency and safety) goals, rather than providing a large amount of low-level information.

- Even though the matter of alarm acknowledgement and handling has been discussed previously, extra care must be taken to ensure that visual alarms can be seen from everywhere they need to be seen in the control room or elsewhere in the facility (e.g., on local field panels). Further, in case visual contact with the alarming functionality is insufficient, there should be auditory alarming as well.

- Depending on the organization's policy regarding the use of procedures and other operating documentation (e.g., mandatory or not), ensure that documents are easy to reach, consult, and put away. Often, documents are on a carousel or on shelves (computer-based documents are not discussed here as their use should be addressed as part of detailed display design). If they are to be used to respond to an alarm, to access emergency procedures, or to support day-to-day operations, then each minute spent fetching or returning a document is a minute not spent working on the process. Ensure that they are handy, without restricting access to workstations.

- Printers are commonly used to display information offline. Unfortunately, they tend to be noisy and this often becomes quite noticeable when the rest of the plant is quiet. This forces the staff to speak more loudly, and quickly becomes a source of irritation. This can be dealt with by acquiring quiet printers, by surrounding them with acoustic enclosures, or by relocating them to an area where they will not contribute to the noise level in the room.

As discussed above, there are several rules and principles that govern the organization of individual workstations and other HMI components. Those rules and principles partly address the relationships between the components but they are not sufficient to fully organize them. To do so, we need to have a method that will help determine how the HMI components should be organized with respect to one another. The next section discusses such a method.

Locating the HMI Components

It is important to select the best location within a given work area for the various HMI components in order to minimize hindrance to the operating staff in carrying out their tasks. For example, operating manuals and procedures should be placed near the location where they are expected to be used; this may be near the operator's desk, but if several operators need to use the same information, then a more centrally accessible area may be more suitable. Another example may be hardwired, backup, workstations or even just sources of ancillary information (e.g., logbooks) that should be readily accessible to the operator in case they are needed. Since we may need to satisfy competing needs when locating those HMI components, it is desirable to use a method to help in doing so.

Several such methods exist that could be used[2]. The one that we will discuss here is especially useful because it is simple to understand and use; the method is known as *link analysis*. Briefly, this method entails identifying the strength of the relationship between all pairs of entities. The type of relationship to consider will depend on the situation. In many cases, it is appropriate to consider the need for communication (e.g., how many times per shift will the operator use the procedures, or phone, or how many times do operators need to work together to coordinate maneuvers?). Having defined the type of relationship that we wish to consider, we need to quantify it. For example, let's assume that we wish to organize the HMI components in the control room so as to minimize the traveling that the operators will have to do between them. We would first list all of the relevant HMI components. Then, we would quantify the strength of each of the relationships between all of those components; here, this corresponds to the number of moves that the operators will normally make between those components, for example over the duration of a shift. If we wish for our analysis to be complete, we need to account for all of the situations of interest such as normal operations, start-up, shutdowns, management of incidents, etc. The best way to collect information on the frequency of those moves is through direct observation. An alternate means to collect this information is to obtain operator's opinions, but beware that this type of subjective information can be inaccurate.

---

2. Finding appropriate arrangements for components, production equipment, etc., is often termed layout planning and is usually achieved by using methods drawn from industrial engineering. Several sophisticated methods exist such as those discussed in *Systematic Layout Planning (Muther and Wheeler, 1994);* some software-based algorithms are also available.

Once the relationships have been defined and quantified, a simple graphic technique can be used to optimize the location of the components.

Figures 8-2 to 8-4 show a simple example for a fictitious case involving two operator desks, one operating procedures carousel and one breaker panel. In this example, we wish to optimize the organization with respect to the number of moves that the operator has to make between the HMI components:

- Figure 8-2 is the number of times (perhaps averaged over several shifts) that an individual had to move from one location to another;

- Figure 8-3 is a bilateral summary of the number of moves between each pair of locations, with a line (or link) thickness commensurate with the number of bilateral moves; and

- Figure 8-4 is the resulting link analysis graph, before and after optimization. Optimization consists in moving the entities that have a strong relationship (as evidenced by line thickness) closer together while being free to keep the entities that have a weak relationship at a distance.

Although there is some subjectivity involved in the method, it is quite useful and often allows one to achieve a good organization of the HMI components. A key factor in successfully applying the method is to properly identify and quantify the various relationships. On this matter, it is important to use data that is as representative as possible. For example, if one is interested in minimizing the number of moves required for the operator, then a sampling of the number of moves should be carried out in different, and representative, situations and weighed appropriately. A description of the link analysis method can be found in (Kirwan & Ainsworth, 1992) as well as in many industrial engineering handbooks.

Even when using a method such as link analysis to organize the HMI components, and in keeping with the iterative approach that we have advocated all along, it is extremely important to ensure that the arrived upon solution will in fact satisfy the needs of the operators and of the organization. Therefore, creating and validating a mock-up should be considered an indispensable step in the process. This needs not be difficult, and simple techniques can be used in this type of situation. For example,

Locating the HMI Components

	Operator 1 Desk	Operator 2 Desk	Operating Procedures Carousel	Breaker Panel
Operator 1 Desk		4	9	0
Operator 2 Desk	11		2	6
Operating Procedures Carousel	3	2		1
Breaker Panel	1	4	1	

Figure 8–2 *Raw Data—Moves*

Interactions	No.	Line
Operator 1 Desk < = > Operator 2 Desk	15	▮
Operator 1 Desk < = > Operating Procedures Carousel	12	▮
Operator 1 Desk < = > Breaker Panel	1	│
Operator 2 Desk < = > Operating Procedures Carousel	4	│
Operator 2 Desk < = > Breaker Panel	10	▮
Operating Procedures Carousel < = > Breaker Panel	2	│

Figure 8–3 *Summary of Moves—Bilateral*

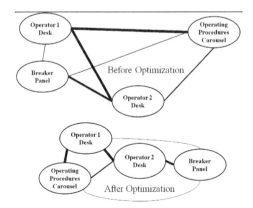

Figure 8–4 *Results from Analysis and Optimization*

using objects of a size similar to the projected consoles often yields a good sense of the dimensions and clearances available than can be obtained from an engineering drawing. Alternatively, one can use masking tape on the floor to delimit the footprints of consoles and panels; this is very cost-effective and is extremely convincing to those observing the test of the layout.

Transition

In this chapter, we addressed how HMIs are generally organized in a control room, and provided some guidance on how to maximize the operator's span of monitoring and control.

As we have seen, HMIs used in a plant would include, by definition, all interfaces between operators and information required for monitoring and controlling the plant. Two such types of interface have gained prominence in the process industry in terms of their potential role in enhancing safety and productivity: operating procedures and decision support systems. They will be the subject of the next chapter.

References and Additional Readings

The following are references as well as additional sources of information readers may find useful.

Paper-based Publications

Kirwan, B. and Ainsworth, L.K. (Eds) 1992. *A Guide to Task Analysis*, Taylor and Francis, London, 417 p.

Mil-Std 1472F Design Criteria—Human Engineering contains several sections relevant to console design.

Muther, R., Wheeler, J.D., Simplified Systematic Layout Planning, ISBN: 0933684-09-6, ISBN-13: 978-0-933684-09-6, Third Edition 1994, 35 pages. This is, according to the publisher, a simplified version of the original *Systematic Layout Planning* method originated by Muther.

The aforementioned standard, ISO 11064 Ergonomic design of control centres, addresses the design of control centers and control rooms. This standard is made up of:

- ISO 11064-1:2000 Ergonomic design of control centres—Part 1: Principles for the design of control centres

- ISO 11064-2:2000 Ergonomic design of control centres—Part 2: Principles for the arrangement of control suites

- ISO 11064-3:1999 Ergonomic design of control centres—Part 3: Control room layout

- ISO 11064-4:2004 Ergonomic design of control centres—Part 4: Layout and dimensions of workstations

- ISO 11064-6:2005 Ergonomic design of control centres—Part 6: Environmental requirements for control centres

- ISO 11064-7:2006 Ergonomic design of control centres—Part 7: Principles for the evaluation of control centres

This standard contains a great deal of up-to-date information on how to best design and validate a control center and control room, including a process, in Part 3, that can be followed to achieve this. Unsurprisingly, the process is iterative and assessment plays a large role in it.

ISA-RP60.3-1985 Human Engineering for Control Centers provides recommended practices aimed at taking into account human characteristics in designing control rooms. The document dates back some years, but it contains concise information on anthropometrics and general design guidance, as well as environmental information. It could be used for initial guidance.

Lastly, and in spite of its "office" flavor, the ISO 9241 Ergonomic requirements for office work with visual display terminals (VDTs) series of standards provides useful guidance. Aspects such as keyboards, displays, etc., are addressed.

Web-based Resources

Chapter 10 in the Federal Aviation Administration's (FAA's) "Human Factors Design Standard (HFDS)" is available at *http://hf.tc.faa.gov/hfds/default.htm* and contains relevant sources of guidance for workstation and workplace design.

The site of the Canadian Centre for Occupational Health and Safety provides a great deal of guidance on office ergonomics. It is available at *http://www.ccohs.ca/oshanswers/ergonomics/office/* and contains information relevant to the design of computerized workstations.

9

Additional HMI Components

We earlier recognized that an HMI includes not only the panels, be they computer-based or hardwired, that the operator can see and touch, but also all of the means by which he or she can obtain information, make decisions, or carry out control actions.

This chapter discusses two additional components that form part of the HMI. First, we will examine operating procedures and how they can best be produced and integrated into the rest of the HMI. Note that we will address paper-based as well as electronic or computerized procedures. While it may seem unusual to discuss procedures as part of HMI design, this reflects a modern trend towards recognizing the role that procedures play in supporting operator decision-making. It also recognizes the transition to more procedures being displayed using computers; they are thus part of display design. We will then look at decision-support systems and the challenges they pose and present some principles for incorporating them into a successful HMI.

Operating Procedures

Operating procedures have become more popular and have even become mandatory in some domains in the last few years. It can reasonably be expected that this trend will continue under the influence of several forces, including regulatory pressures, quality management systems, and even the sheer benefits of systematizing operations. This then creates the need to determine how to best integrate procedures into the overall HMI.

The Benefits of Procedures

Before looking at how to integrate procedures into HMIs it is useful to try to understand the benefits, other than complying with regulations, which procedures can provide. There are several benefits, and understanding these will help tremendously, not only in complying with the various rules that mandate the use of operating procedures, but also in understanding how procedures can make business sense when used properly. A key message here is to consider the use of procedures as a powerful tool to improve process performance, rather than considering them as just another task required to satisfy regulators or quality assurance people.

The first benefit brought about by using procedures is the systematization of operating practices, which, when thoughtfully carried out, ensures that everybody uses the same and hopefully the best methods to monitor, diagnose, and control the process. Not only should this result in fewer process disruptions and upsets, but process safety should also benefit.

Another benefit is to stabilize the behavior of the operator in difficult or stressful situations. From our earlier discussion on human decision-making, you may recall that humans work best and most reliably when reusing a known successful method, rather than trying to improvise a new one each time they do something. Our working universe is replete with relevant examples. While this is somewhat humbling, we humans are just not that good at creating new and flawless methods, especially in an incident or accident situation. A well thought out set of operating procedures can serve to ensure that operator response to specified and significant process disturbances takes the form of steps that are based on the best collective judgment of operation and technical staff. Good procedures can also help alleviate the anxiety an operator could feel if he or she had to improvise a series of steps to prevent the escalation of a crisis in a complex production unit.

With heightened visibility being given to succession planning in several industries, procedures have also become one of the most useful tools for ensuring that corporate knowledge gets passed on to new employees. Not only are procedures useful in the operation process, they also act as a knowledge repository and as training tools.

With these benefits in mind, we will look at a process that can be used to develop procedures. Note that for all practical purposes, the same process can be used to develop either paper-based or electronic procedures.

The Development Process

There is no set cross-industry standard for writing operating procedures, but there are a number of generic steps that should be followed to ensure that the resulting procedures are appropriate. The following sections provide guidance on how to best carry out those steps. In addition, there are several sources of detailed guidance available (see the *References and Additional Readings* section at the end of this chapter).

Define the Purpose and Scope of the Procedures

The first and most important step is to decide for which situations you will create procedures. It is neither practical nor necessary to create procedures for every single task; creating unnecessary procedures will hamper rather than help your efforts. Since it takes a significant commitment of time and effort to produce useful written procedures, and since not all procedures need to undergo the same level of scrutiny regarding their production and validation, it is often a good idea to distinguish between the tasks for which procedures will be developed. Those might include:

- Normal situations, such as starting up or shutting down the process, making product changes, or carrying out other tasks as they are normally carried out.

- Those cases where the operator has to cope with incidents or abnormal situations, such as what to do if equipment A is unavailable or if a discharge to the environment is to be prevented.

- The situations where the operator has to diagnose the cause(s) of a malfunction or process upset.

Apart from the above classical types of procedures, it is also often a good idea to have a set of procedures to be used for commissioning. Even though they may only be used once, they may well prove their worth during these high-stress, high-risk situations. Finally, and even though they are not strictly procedures, a set of references is often useful—if not for day-to-day operation, then at least to provide more in-depth knowledge to the operator and to support training.

Defining Training Requirements for Using the Procedures

There is often a perception that it is only a matter of being able to read to be able to follow a procedure, but this is wrong. There is a great deal of implied knowledge required to follow a seemingly simple procedure. For example, as a hypothetical "get the car out of the driveway" procedure, the step "start the engine" requires the would-be driver to know how to use the key, where to get it, and how to determine when the engine has actually started. Further, the driver also needs to know that after several unsuccessful attempts, he or she needs to get the car serviced, as there is no use in draining the battery. Therefore, and at minimum, knowledge and training requirements must be ascertained in enough detail that appropriate content can be provided in the procedure. A good link between procedures and training often helps keep procedures shorter and more straightforward, since some of the detailed knowledge required can best be conveyed through basic training.

Specifying How to Write, Validate and Verify Procedures

Two complementary aspects of procedure writing need to be addressed first:

- Programmatic requirements determine who will write which procedures, based on what information, and who will review and approve their content. Normally, the writers and reviewer(s) will be technically competent in the details of the operation covered by the procedure, while the approver will normally declare whether the procedure has been developed in accordance with the organization's requirements. You can establish a set of policies and procedures that will cover these topics; this will help you keep the process under control and will ease deploying the finished product. Note that deciding who will write the procedures is often difficult. Procedure writers often lack the technical or operation knowledge required, while operators often lack technical writing skills (and sometimes the inclination for producing documents). A good approach is to use a mixture of both, where the technical writer will elicit knowledge from operation subject matter experts (SMEs) and technical specialists. This minimizes the use of scarce operation resources while ensuring that good quality documents are produced.

- Requirements on how to write the procedures where we will specify the vocabulary to be used, how to write a specific step (IF, THEN, etc.), which header and footer to use, and which font to use. Additional guidance on reducing the complexity of the procedure can also be provided. These requirements are often embodied in a writer's guide, which can be custom-made for a specific organization or adapted from a commercial publication (several good manuals on technical writing are available). If you recall our discussions on HMI design, you will realize that this writer's guide is, in fact, the equivalent of a style guide for display design.

Once you have the required guidance for controlling the process and producing the content, you need to determine how you will validate and verify your procedures. Skipping these steps will essentially negate any benefits that you might have achieved since an invalidated or unverified procedure is about as reliable as an untested new design for a parachute.

It is, however, a fact of life that resources are limited, and it will not be possible to submit all of the new procedures to the same level of scrutiny regarding their verification and validation. The following general guidelines will help you prioritize your verification and validation assessments:

- Each procedure should be reviewed and verified for technical correctness and for compliance with the writer's guide by staff at least as knowledgeable as the author(s) of the procedure.

- Each procedure should be validated. In the author's experience, even though a procedure may be technically correct, there are often operational inefficiencies or even unexpected potential interpretations due to how the procedure is written, and those may only become obvious during a simulation with an operator. For important or critical procedures, it is worthwhile (and often will more than pay for the validation exercise) to have one or more operators carry out a simulation of the procedure. The preferred means of validating or simulating a procedure is to "walk it down" physically in the plant, going through the motions described by the procedure without necessarily actuating any control. This allows one to assess if the labels on the equipment actually match those in the procedure, or if the controls that are to be operated are accessible, or if the procedure as written takes into account the walking that the operator has to do around the plant to execute the plan. A less powerful approach, which may be used for less important procedures, is to carry out "table top" or "talk through" validations without positively verifying that the content of the procedure matches the equipment and layout of the physical plant. If one decides not to validate a procedure prior to its use, there will be a high degree of risk that its shortcomings show up when it is used for real. Our position is that if it is worth writing a procedure, then it is necessary to ensure that the procedure will work as expected.

Determining the Content of Procedures

As hinted previously, determining the appropriate content will require an iterative approach. This is just another example of the fact that procedures are a subset of the HMI and their design is subject to the same challenges. Determining the content will require that the best description of the operator's task be obtained, that it be validated by engineering or other knowledgeable technical resources, that unavoidable (and welcome) discrepancies between the inputs of various subject matter experts be identified and resolved, and that the resulting content be translated into a clear set of steps.

Just as there are a number of things to consider in connection with the process of producing procedures, there are a number of detailed design principles and rules that should be used. They are addressed in the next section.

Procedure Design Principles and Rules

The objective in this section is to explain, through some examples, how design principles and rules impact procedures writing. There is no attempt here at exhaustiveness. More complete treatments can be found in specialized technical writing guides (see the *References and Additional Readings* section at the end of this chapter).

Structuring with Goals

Structure is important in procedures, as it helps the operator understand the objectives and reduces the likelihood of human error. Further, having a clear objective will help the operator identify and even overcome an erroneous step in a procedure. Here are a few design principles:

- Procedures should incorporate various levels (goals or objectives, tasks, actions) to accommodate operators with varied levels of qualification.

- Determine the required level of detail for a procedure, depending on the type of procedure, its frequency of use, and the experience levels of the operators.

- Link the operating goals, the actions required, the actual process situation, and the expected process evolution.

Complying with and Deviating from Procedure Steps

No matter how diligent the procedure writer, there will be occasions where the operator will see a need to deviate from the procedure. Examples of deviations include changing the order of the steps in the procedure, skipping altogether a series of steps, or even introducing new steps. How deviations from the procedure are to be managed must be determined before the procedure is actually deployed or even produced. Here are some guidelines:

- Define rules governing whether the procedures must be followed strictly as written, or if some leeway is permitted. This determination will be normally driven by the risk involved with the task, i.e., a higher risk will usually call for a stricter level of compliance. In some quality-driven processes, it may be mandated that a change cannot be made to the process until the related procedure has been reviewed and approved.

- Be careful not to make compliance requirements more strict than absolutely needed, as this may increase the cost of producing or executing the procedures. On the other hand, make the requirements as strict as necessary, considering the risks involved.

- Define a process to handle non-intent changes—such as editorial or other minor changes that do not detract from the procedure's intent—which occur during operations.

- Define a process for promptly reviewing and authorizing, as appropriate, intent changes that occur during operations. This will normally require a higher level of scrutiny than will non-intent changes.

You should also define a process that allows operating or other staff to submit enhancements to existing procedures.

Writing Procedure Steps

Producing the actual procedure differs from usual technical writing in several respects. You should think of a procedure as a "program" that an operator will carry out; there is thus a premium put on writing the steps in a way that will prevent any ambiguity in understanding and carrying them out. It is also important that any extraneous verbiage be avoided since the document will usually be used when the operator is trying to carry out a set of specific tasks.

To help the procedure writer embody best practices in the actual procedural steps, detailed design rules are available in the scientific and technical literature. Those rules cover factors impacting the legibility and comprehensibility of those steps, as well as ensuring that the correct version of a procedure is used. The Appendix contains a number of those rules.

Considerations for Computerized Procedures

Computerized procedures offer several advantages over paper-based procedures. For example, it is possible to determine automatically if the entry conditions of a procedure have been met, to display information supporting the operator in making a decision about one of the steps in the procedure, or even to offer control functionalities on screen, right next to the step where the action is required.

In spite of those advantages, there are several challenges in implementing effective computerized procedures. For example, navigation within a computerized procedure is often more difficult than navigation through paper-based procedures, as several of the usual physical cues associated with paper (e.g., thickness, ease of putting in placeholders) are missing. There are various levels of automation associated with computerized procedures, ranging from static text on a computer display to a fully interactive, automated procedure that allows one to read process variables, or carry out control actions. Sophisticated computerized procedure systems that provide support in choosing the appropriate procedure from a set of potential procedures, also exist.

While it is beyond the scope of this discussion to provide guidance on how to design electronic procedures, we will still briefly mention some key principles as computerized procedures form part of a growing number of HMIs:

- When computerized procedures are used, there should be a back-up system in case the primary system fails. This is especially important if the procedures are used to manage incidents or accidents.

- It is important to recognize that computerized procedures are just another type of display. In fact, in many cases where a display has been designed to support a specific task (e.g., a display dedicated to starting up a system that is made up of several components), the display in effect embodies the elements of a procedure. Thus, to the extent practical, the same conventions should be used for the computerized procedures as for the rest of the HMI. When some ambiguity is likely (e.g., checkboxes are used to mark a choice in the HMI, but they are often used as markers of progression in computerized procedures), the appearance of the components for which there might be ambiguity should be made visually different.

- If several computerized procedures are available, they should be grouped logically and be easily accessible.

- There are design choices to make when designing automated procedures systems regarding the amount of control that the operator will have on the pace of work and the ability that he will have to override decisions made by the computer (e.g., automatic branching to a given part in the procedure); in some cases, the computer may also select which procedure should be used. It is recommended that the user should retain the ultimate choice as to pace, selection of path through the procedure and choice of the appropriate procedure. There are several reasons for this: the user may have information that the computerized system does not have, or the computerized system may have a malfunctioning sensor, or the procedure may simply be inappropriate for the task at hand. Of course, there is always the possibility that the operator is wrong. While there is no guaranteed error-free method for choosing between the operator's opinion and the computerized system's choice, a clear policy on who has the last word (and ultimate responsibility) coupled with training on the capabilities and limits of the computerized system will often provide an adequate solution.

We are fortunate that guidance on computerized procedures has been generated in other industries (e.g., nuclear, aerospace) and that it can be used with very few changes. The section *References and Additional Readings* provides pointers as to where this additional guidance can be found. Also, note that several of the detailed design rules provided for paper-based procedures are also useful for computerized procedures with few, if any, adaptations.

Integration with the HMI

Having completed our discussion on how to organize and structure procedures, we need to determine how they are going to be integrated with the rest of the HMI design. It now becomes obvious why we treat procedures as part of the overall HMI, since they are used concurrently with the other means of data and information gathering and control execution that are available to the operator. While other HMI components provide operator information and control functionalities, procedures provide guidance to the operators on how to carry out their tasks. In fact, if displays have been designed while taking into account the task model of the operator, as has been explained during our discussion on HMI design, the displays themselves can be seen as a kind of procedure.

As an example, let's assume you have designed a computer-based HMI using the methods and processes that have been presented before; let's further assume you have produced the procedures while taking into account the processes, principles, and rules outlined in the previous sections. Here are some steps that will facilitate the concurrent use of the procedures with the HMI to best support the operator:

- Terms used in designing the HMI and writing the procedures (e.g., valve or controller designation, units, equipment or process system naming) should be the same. This is best handled by specifying the source for these terms in the HMI design style guide and in the procedure writing guide.

- It is important to maintain continuity between the information and control functionality provided by the HMI and the task guidance provided by the procedures. A good example of this is an alarm response procedure. In some applications, paper-based procedures provide a standardized operator response to alarms generated by the computer-based system. For these cases, the alarm designation provided by displays should be the same as that used for the paper-based alarm-handling procedures. These, in turn, should use the same terminology as the ones found on the computerized displays.

- The procedures and the HMI should undergo validation simultaneously. If a procedure refers to a display, or to a hardwired panel, it should be verified and validated through a simulation involving the use of the display or panel. Similarly, if a display or panel is modified, the corresponding procedure should be used when validating the display or panel.

- There should be a mechanism to ensure that, whenever a change is made to either the procedures or the HMI, a cross-check is performed to keep these components in sync with one another. This is another example of the inter-dependence between the HMI and the procedures.

Procedures constitute one of the most cost-effective means to enhance the safety and productivity of a plant. Their integration with the rest of the HMI, while sometimes difficult, is an important step toward smoother operations.

As a last comment on this topic, we have only touched on the usual use of procedures. Be aware that producing procedures goes beyond the mere generation of documents, since procedures essentially embody the incident response strategy of an organization. Defining an appropriate set of procedures may thus have significant positive impacts on the operational efficiency of a plant if it is used as a means to implement optimal event recovery strategies. Further, there are more fundamental matters that should be considered if one wants to take full advantage of incident management. For example, research in the nuclear sector has identified two main classes of procedures: event-based and symptom-based. While the former corresponds to the procedures commonly used in process plants to respond to specified situations, the latter provides a generic and powerful response to unanticipated events. These two classes of procedures are typically used simultaneously to provide comprehensive coverage for events that can occur in a plant. This topic, however, goes beyond our objectives for this book.

Decision-Support Systems

The term decision-support system (DSS) is a bit of a misnomer, since just about any piece of information made available to an operator can technically be called that; in fact, operating procedures are probably the most widely used, and one of the most effective, decision-support systems available to the operator. However, to use modern parlance and at the expense of purity of the language, decision-support systems have been more commonly introduced with the advent of cheaper and more powerful computer systems. DSSs come in all shapes and forms, and typically include expert systems, configuration management systems, alarm support systems, etc. As such, they can be used to assist the operator in detecting, and sometimes managing, situations that may be difficult, and that are poorly supported by the existing HMI.

In this section, we will examine some challenges associated with the use of DSSs, and we will discuss how to ease their integration and simultaneous use with the rest of the HMI.

Challenges

There are about as many challenges associated with the use of DSSs as there are different types of DSSs. Understanding these challenges will help us take better advantage of these systems.

A first challenge has to do with the fact that DSSs are often add-ons to existing systems. For example, a system meant to help the operator better handle alarm messages may be added to a control room suite of displays and panels. This means that, when an alarm occurs, say on a digital control computer (DCC), the operator would have to turn to access the information and respond to the alarm. This may involve using a different keyboard and pointing device, terminology, presentation conventions, and navigation scheme, thus potentially imposing more work on the operator. Here we can get a first glimpse of one of the rules for successfully integrating a DSS with the HMI—that is, it must provide a greater benefit to the operator than its cost in terms of operator workload, if we want it to be used.

Another challenge relates to the technologies used by various DSSs. For example, expert systems use rules (IF ... THEN ... constructs) that allow the DSS to reason from a set of facts to new or inferred facts, mimicking some of the steps of human decision-making. However, there are some difficulties with this type of reasoning and with the system's interaction with the operator. For one thing, the set of rules cannot be guaranteed to be complete, casting doubt on the competence of a rule-based DSS. This is often not catastrophic, as the purpose is to help the operator rather than to be able to solve every possible case. However, this implies that there must be a means to identify and cope with situations that are beyond the competence of the DSS. Also, the operator cannot be expected to blindly follow the advice provided by the DSS. The ability of the latter to explain itself is thus very important. However, as we have seen, the reasoning used by a rule-based system is simply the triggering of a series of rules leading from a set of facts to some logical conclusion. This often makes the explanation somewhat obscure to the operator.

The last issue we will discuss is the need for low-level integration between the DSS and the rest of the HMI, both in terms of the content and the hardware used. To be truly useful, the DSS must be designed with the overall mission of the human-machine system in mind, rather than with the main focus being on the performance of the DSS itself. Further, since the DSS is frequently an add-on, it must not increase the normal workload of the operator, or if it does, it must provide enough benefits to justify the increase. Lastly, since you have put a great deal of effort into the design of the HMIs and the procedures, the DSS must reinforce the benefits of those components rather than come into opposition with them.

This set of challenges is not complete, but is sufficient to suggest some principles for integrating the DSS with the rest of the HMI and to optimize the joint operation of both.

Integration into the HMI

Our previous discussion allows us to define some principles for getting the most out of a DSS:

- There must be a clear and unambiguous policy for using the DSS as part of the portion of operations for which it is intended. In particular, should there be a situation where the DSS goes against the best judgment of the operator, or against the information provided by the rest of the HMI, there should be clear guidelines as to where the decision-making authority lies.

- It is often a good idea to first introduce a DSS through the training simulator (if one is used for training); this allows the operator to gain experience with the system and, more importantly, to determine the boundaries of its competence. Also, future enhancements or fixes to the system can be validated on this version before they are transitioned to the actual plant.

- The use and competence of the DSS, and the policy for using it, should be incorporated into the new operator training program; training should also be provided to existing operators.

- The DSS has to pay its way in the control room, or in the plant, in terms of the balance between its contribution to the safe and productive operation of the plant and the changes it requires to operating practices and the work area.

- The DSS should use a style guide similar to the one used for the rest of the HMI; ideally, it should be the same style guide.

Further, the terminology used should be the same as that used in the HMI and in the procedures.

- Communication between the DSS and the operator, including explanations for recommendations on decisions, must be in a language that corresponds to the one used by the operator. For example, advice and explanations should be offered in terms of the evolution of variables in trends, components in piping and instrumentation diagrams, etc., rather than in textual outputs and reference to reasoning rules.

- Minimize the type and number of interaction devices, as there is often a tendency to introduce new hardware when a DSS is provided. For example, because desk space is usually limited and crowded, every effort should be made to avoid adding a new keyboard and a new mouse, or even a new monitor, when a new DSS is introduced.

DSSs are a relatively new ingredient in the human-machine system, and they are still being actively researched. It should be expected that additional guidance will be forthcoming as we learn more about their optimal blending with an HMI.

Conclusion

In this chapter, we have discussed how to leverage operating knowledge captured in procedures and, to a lesser extent, in decision-support systems. We can expect that those types of approaches will play an increasing role in supporting the operators in the years to come, because of regulatory pressures (e.g., to minimize risks to the environment), to enhance knowledge sharing amongst personnel or simply to systematize the use of good operating practices.

This chapter concludes our discussion of the design of human-machine interfaces for process control applications. As we have seen, it is necessary to master several techniques and concepts, as well as to apply the appropriate design principles and rules, to design successful human-machine interfaces. This is worthwhile, however, as designing better human-machine systems provides one of the easiest paths towards the improvement of safe and efficient monitoring and control of industrial processes.

Paper-based Publications

CAP 708, Guidance on the Design, Presentation and Use of Electronic Checklists, Civil Aviation Authority, ISBN 0 86039 804 8, First edition December 2000, Reprinted March 2005. A fairly recent and comprehensive document on the design of computerized procedures (which they call electronic checklists in aviation).

Human-System Interface Design Review Guidelines, - NUREG-0700, Rev. 2. In Section 8, there is quite extensive guidance on how to design computerized procedures.

Procedure Writing: Principles and Practices, Douglas Wieringa, Christopher J. Moore, and Valerie Elizabeth Barnes, Battelle Press, 1993, 211 pages. This is a well-written, generic, and comprehensive source of information on how to best write procedures. It also provides a fair bit of information on why certain forms of writing are better than others. Suited for paper-based procedures.

Writing Operating Procedures for Process Plants, Southwestern Books, 1995, 302 pages, by Ian Sutton. A clear, practical and "how-to" oriented book, suited for the process industry. One of its strengths is to provide guidance on how to organize and structure procedures. Suited for paper-based procedures.

Design Guidelines and Components

This summarizes detailed design guidelines on specific HMI components. The information provided is not exhaustive, but it provides a concise source of information in a convenient way. It has been compiled from a variety of sources and represents a number of good practices. Guidance is provided for:

Computer-based HMIs

- General characteristics
  - Visual coding and conventions
  - Watchdog timer
  - Sound
  - Legibility
  - Annunciation and alarms
- Windows
  - Main
  - Secondary or pop-up
- Display components
  - Graphics (trends, bars)
  - Digital displays
  - Alarm message lists

- Inputs
  - Touchscreens
  - Joysticks
  - Mouse

Hardwired panels

- General
  - Workstations
- Controls
  - Buttons
  - Knobs
- Displays
  - Meters
  - Chart recorders

Large screen displays (LSDs)

Procedures

One last word of caution is in order. Even though reasonable care has been taken in compiling this guidance, it is important to remember that any design rule or guidance needs to be considered within its context of use. A rule that makes perfect sense in a given situation may yield bad results in another; further, it may be necessary, in some situations, to bend some rules to achieve a better overall design.

Table A–1   Computer-based HMI

Characteristic	When to use	Recommendations
Visual coding and conventions	■ A visual coding method should be used consistently throughout the interface.	■ Visual indicators should be coded by color, size, location, shape, or flash coding as applicable.
Color	■ Color should be used to aid in identifying various keys by function or use. ■ Color coding may be employed to differentiate between classes of information in complex, dense, or critical displays. ■ Color can be used to help the user identify the relevant information.	■ Ensure that the display design can be used without any color, then add color to enhance it. ■ Define a convention for the use of colors and use it throughout the interface. ■ Use secondary coding to cater to the needs of the color-blind. ■ Red should be reserved for emergency functions. ■ Color-filled symbols should be used instead of color-outlined symbols. ■ The number of colors used for coding should be kept to the minimum needed for providing sufficient information.
Shape (icons or symbols)	■ When secondary coding is necessary to support identification. ■ Shape coding may be used for search and identification tasks.	■ When possible, icons should be simple drawings that suggest the physical object or operation they represent. Icons are frequently too small and filled with illegible details. ■ A symbol should be an analog of the object it represents. ■ When shape coding is used, the codes selected should be based on established standards or conventional meanings.
Flashing	■ To draw attention to information that requires prompt response.	■ Use flashing with extreme parsimony as it gets very annoying to the operator. ■ Flash rate should be between 3 to 5 Hz, with a 50% duty cycle. ■ If several objects are flashing, their flashing should be synchronized. ■ There should be an easy means for the operator to acknowledge the situation and stop the flashing.

Table A-1   Computer-based HMI (cont'd)

Characteristic	When to use	Recommendations
Stale data (or unrefreshed data)	■ To identify situations where the data source (e.g., the network, the sensor, or transmitter) is no longer updated or available.	■ A clear indicator should be used to indicate that a piece of data is no longer updated. A good way is to fill the field with "*****". ■ Avoid displaying the old value as the operator may believe that the data is still valid. If the old value is of any use, give access to it using other means.
Flow (mass, energy, information)	■ To arrange components on a display.	■ The flow of mass, energy, or even information should normally be from left to right, or from top to bottom.
Watchdog timer	■ To indicate that the computerized system is no longer available or serviceable.	■ Use an explicit indicator to the effect that the computerized system is no longer available. For example, open a window stating this. This can be implemented using off-the-shelf components. ■ Avoid using *implicit* indicators (e.g., using the clock and expecting that the operator will notice that the time is no longer incrementing – this is a very error-prone implementation).
Sound	■ An auditory signal should be used to alert the user of any condition of which the user must be made immediately aware.	■ Each audio signal should be unambiguous and easily distinguishable from every other tone in the control room. ■ The signal intensity should be such that users can reliably discern the signal above the ambient control room noise. The signal must be present where the operator is expected to be. ■ Audio signals should be designed to minimize irritation and startle. ■ There must be a convenient way to silence the sound once the operator is aware of the alarm.

Table A–1 Computer-based HMI (cont'd)

Characteristic	When to use	Recommendations
Legibility (readability)	■ For any labeling purposes.	■ Ensure that the data is readable from the location where it is expected to be read–typically for "desktop applications" at distance of 75 cm. Quite often, though, the distance may be greater, so you should look at the actual work station. ■ Character height should subtend a visual angle of 16 minutes as a minimum; a visual angle of 20 or 22 minutes is preferred. For example: | Distance	Visual angle, 16 minutes	Visual angle, 22 minutes
72 cm	3,4 mm	4,6 mm
144 cm	6,7 mm	9,2 mm
288 cm	13,4 mm	18,4 mm | ■ Letter height should be identical for all labels within the same hierarchical level, based on the maximum viewing distance. ■ Labels should be in capital letters. ■ The minimum space between words should be one character width. ■ The minimum space between lines should be one-half of the character height.

Table A–1 Computer-based HMI (cont'd)

Characteristic	When to use	Recommendations
Annunciation and alarms	■ When providing alarm or other annunciation	■ The annunciation for status changes (e.g., pump XYZ running or shutdown) should differ from the annunciation for alarm information. For example, a different list of messages should be used for alarms and for status information; discrete alarm and status indicators should have a different presentation. The objective is to prevent data flooding to the operator. ■ The same visual coding for annunciation and alarms should be used throughout the displays. ■ There should be an easy means for the operator to determine if an alarm is inhibited. This could be through a list with some redundant indication on displays.

Table A-2 Windows

Characteristic	When to use	Recommendations
Main	■ A main window, or series of main windows, can be found on most computerized HMIs.	*(diagram showing a window with Menu, Window Controls, File View Tools Reports Help, Toolbar, and Title labels)*

- ■ The menu should be defined in accordance with the guideline provided in the style guide for the selected application. The grouping of the items under a given menu should also be done in accordance with the style guide.
- ■ The tools on the toolbar provide shortcuts to frequently-used functions. Every tool should have an equivalent menu item.
- ■ To minimize the need for the operator to manage windows, windows controls (e.g., minimize, maximize) should be used very sparingly and, if possible, not at all.
- ■ There is often a recommendation to minimize the number of pixels used to display information (this is often known as the density of the information rule). However, operators who use displays on a frequent basis tend to prefer a high density of information on a display; further, high-density displays reduce the number of windows used and the need to navigate. On the whole, we recommend a higher information density for displays that will be frequently used by well-trained operators.

Table A–2 Windows (cont'd)

Characteristic	When to use	Recommendations
Secondary (sometimes called pop-up)	■ To provide additional or complimentary information. ■ To provide control functionality (e.g., for an actuator). ■ To display information. ■ To display exceptions (an exception is usually a system generated message that is not process related).	*[Diagram showing a secondary window with Description, Window Controls, and OK/Cancel buttons]* Secondary windows should: ■ Open in the same area of the display. ■ Have the same size; if necessary, standardize a few sizes that will be re-used. Designers often make the secondary or pop-up windows too small; make them large enough to provide the designer with the surface required to do proper design. ■ Be uniquely identified. ■ Avoid allowing the operator to resize the windows controls (e.g., minimize, maximize) unless there is a compelling need to do so as a minimized secondary window can easily be "lost". ■ Not be superimposed on top of one another. A portion of the population of users has difficulties distinguishing superimposed windows. ■ Use buttons (here, OK and Cancel are only examples) that follow the rules for labeling and locations specified in the style guide. For example: ● OK will normally submit information and close the window.

Table A–2 Windows (cont'd)

Characteristic	When to use	Recommendations
Secondary (continued)		• Apply will submit information but keep the window open. • Cancel will not submit the information but will close the window • Close will simply close a display-only window. ■ Have the focus by default (i.e., which button would be triggered if the operator pressed Enter) always set on the button that leads to the least damageable consequence. If a secondary window is used to display messages from the computerized system (or exceptions), then: ■ In the window, explain what the exception is in plain language (avoid messages like "Error 355 occurred"). ■ Provide an explanation as to the cause of the error. ■ Recommend a course of action to the operator. Secondary windows can be modal or modeless. The difference is that a modal window takes control of the interaction and that the operator must complete the interaction with this window before being able to access the rest of the interface. This may be useful if one wants to force the operator to complete a task before leaving a window. However, beware that modal windows should be used with parsimony as they make the rest of the HMI inaccessible until they are closed.

Table A–3 Display Components

Component	When to use	Recommendations
Graphics (analog displays): general	■ To support the user in the detection of changes and deviations from the norm. ■ When examining and interpreting patterns in numerical data are needed. ■ When two or more continuous variables need to be followed.	■ Graphics should be consistent in design, format, and labeling throughout an application and related applications. ■ For increased needed precision of values, the actual data values should be displayed in addition to the plotted data. ■ Displayed graphics should be clearly labeled.
Trend graphs	■ When a record of past values is needed over a period of time. ■ To display the evolution of a continuous variable. ■ To examine relationships between variables (also shown on trend graphs).	■ Trend displays should be capable of showing data collected during time intervals of different lengths. ■ When the user must compare data represented by separate curves, the curves should be displayed in one combined graph. ■ It should be possible to look at previous data points. ■ The scale should be on the side where the most recent data points are displayed (e.g., if new data points appear on the right hand side of the graph, the scale should be on the right hand side) to help legibility. ■ The current or most recent value should also be displayed numerically. ■ It should be possible to zoom on certain areas of the graphs. ■ When multiple curves are included in a single graph, each curve should be identified directly by an adjacent label. However, it is usually not advisable to display more than four curves on a single graph.

Table A–3   Display Components (cont'd)

Component	When to use	Recommendations
Bar charts (vertical or horizontal)	■ When data must be compared. ■ When single variables need to be viewed over several discrete entities.	■ In a related series of bar charts, a consistent orientation of the bars (vertical or horizontal) should be adopted. ■ If a series of variables has the same nominal value, bar graphs can be used to represent those as a group or as an object, thus facilitating pattern detection by the operator, as shown below. When colors or patterns are used to fill enclosed areas: ■ Color coding should be redundant with another form of coding. ■ Colors should be used on screen, rather than patterning. ■ For printing, simple hatching and shading patterns should be used rather than color. ■ If one bar represents data of particular significance, then that bar should be highlighted.

Table A–3 Display Components (cont'd)

Component	When to use	Recommendations
Digital displays (numeric data)	■ Digital displays should be used for quick, precise readings of quantitative values while trend information is not needed.	■ Numeric values should ordinarily be displayed in the decimal number system. ■ Leading zeros in numeric entries for whole numbers should be suppressed. ■ A number should be displayed as the number of significant digits required by users to perform their tasks. ■ Digital displays should not be updated more than once per second as they may become difficult to read. ■ Numeric displays should accommodate the variable's full range. ■ If users must evaluate the difference between two sets of data, the difference should be presented on the display. ■ All numbers should be oriented upright and read from left to right. ■ Unless ambiguity is probable, it is often not desirable to label the variable; rather, its engineering unit can be used to convey its meaning (e.g., 185 kPa will be interpreted as a pressure, without having to add a label). ■ Group the digits. For example, show 2 578 352 rather than 2578352 to minimize operator error.

Table A–3 Display Components (cont'd)

Component	When to use	Recommendations
Alarm message lists	■ When the current and past status of all the alarms needs to be viewed in the same application.	■ Alarm message lists should be displayed prominently on the displays, preferably on top of the page. ■ The list should allow the presentation of a sufficient number of messages (e.g. sufficient to avoid forcing the operator to navigate or scroll frequently in the list). ■ If priorities can be reliably established, the lists of alarm messages should be segregated by alarm priority, with highest-priority alarms being listed first. ■ Lists of alarm messages should be grouped in relevant operation categories according to function, chronological order or status acknowledgement (active or cleared). ■ Alphanumeric alarm lists should have a blank row, or another delimiter, between every four or five alphanumeric messages. ■ Adding alarm messages to the list should preclude message scrolling. If many alarms are coming in, scrolling makes it difficult to read messages; paging presents an alternative solution. ■ Alphanumeric alarm messages that overflow the first page of alarm messages should be kept on subsequent alarm pages.

Table A-4 Inputs

Component	When to use	Recommendations
Touch screens	■ Touch screen control may be used to provide an overlaying control function to data display devices such as CRTs, dot matrix/segmented displays, electroluminescent displays, programmable indicators, or other display devices where direct visual reference access and optimum direct control access are desired. ■ Tasks involving touch screen use should not require frequent, alternating use of the keyboard.	■ Touch screens should be clearly readable in the work environment. ■ Positive indication of touch-screen actuation should be provided to acknowledge the system response to the control action. ■ System display response time should be less than 100 msec. ■ When used for a critical task, system response should require confirmation. ■ Keys on a touch screen should be regular, symmetrical, and equilateral in shape. If color-coding is used, it should only be a redundant form of coding. ■ For frequent inputs, touch screens should be mounted at 30-45 degrees from the horizontal. Otherwise, fatigue will occur. ■ The designer should • Determine what will be the response of the touch screen if two points are touched simultaneously and then take this into consideration. • Provide a means to de-activate the touch screen for cleaning. • Determine if users will wear gloves when using the touch screen as not all touch screen technologies allow it.

Table A–4 Inputs (cont'd)

Component	When to use	Recommendations
Isotonic or displacement joystick	■ May be used when the task requires precise or continuous control in two or more related dimensions and when positioning accuracy is more critical than positioning speed. ■ May also be used for various display functions such as selecting data from a CRT and the generation of free-drawn graphics.	■ Displacement joysticks used for rate control should be spring-loaded for return to the center when the force is removed. ■ Joysticks should be mounted to provide forearm support. ■ Movement should not exceed 45 degrees from the center position and should be smooth in all directions. ■ Modular devices should be mounted to allow actuation of the joystick without slippage, movement or tilting of the mounting base. ■ Movement should be smooth in all directions.
Isometric joystick	■ When tasks require precise or continuous control movement in two or more related dimensions.	■ The output should be proportional and in the same direction as the applied force. ■ The maximum force for full output should not exceed 118 N or 27 pound-force.

Table A–4 Inputs (cont'd)

Component	When to use	Recommendations
Mouse	■ May be used for data pickoff or for entry of coordinate values.	■ The mouse should be easily movable in any direction without a change of hand grasp and should result in smooth movement of the follower in the same direction ±10 degrees. ■ The mouse should be operable with either the left or the right hand. ■ A complete excursion of the mouse from side-to-side should move the follower from side-to-side on the display. ■ The mouse should have one or more buttons that provide features related to various functions and control actions. ■ Button contact surfaces should be perpendicular to displacement direction and finger motion during actuation. ■ The mouse should be shaped to allow the operator to grasp it with either hand in a relaxed and neutral posture. ■ Ensure that the mouse can be located to the right or to the left of the keyboard to accommodate right- or left-handed users. ■ Right-clicking on a mouse could be used to offer contextual choices to the experienced operator. ■ In that some users are not at ease with the double-click functionality on a mouse, the designer should refrain from requiring actions that are based solely on double-clicking.

Table A–5 Hardwired Panels—General

Component	When to use	Recommendations
Workstations	■ To specify all dimensions of seated or stand up workstations.	■ For detailed dimensions (e.g., height, width, depth, clearances), see Section 5.7 of MIL-STD-1472F, DEPARTMENT OF DEFENSE DESIGN CRITERIA STANDARD - HUMAN ENGINEERING 23 August 1999. This standard can easily be found on the Web.
	■ For stand-up workstations. Status and Annunciation Displays Controls	■ Dedicate a separate zone for the alarms and one for the status information. Both types of information should not be mixed within a given zone to avoid the potential for confusion. ■ Use the dark panel approach where all the annunciation is turned off if there is no alarm. ■ Provide a button for lamp tests and ensure, through administrative procedures, that lamp testing is carried out on a regular basis. You may also want to use high reliability lamps (e.g., with two filaments). ■ Use similar types of devices for similar functions. ■ Ensure that critical controls (e.g., for tripping a system) are adequately accessible, or protected, or both. ■ Ensure that controls cannot, or are unlikely, to be actuated inadvertently.

Table A–6 Hardwired Panels—Controls

Component	When to use	Recommendations
Buttons (knobs)	■ Knobs should be used when low forces or precise adjustments of a continuous variable are required.	■ For most tasks, a moving knob with a fixed scale is preferable. ■ A pointer or marker should be available on the knob when positions of a single revolution must be distinguished.
Rotary knobs		■ Rotary controls should turn clockwise to cause an increase in parameter value with an associated display movement. ■ If there is a time lag between control actuation and system state, there should be an immediate feedback indication of the process and direction of parameter change in real time. ■ Controls should provide the capability to easily affect the parameter controlled, with the required level of precision. ■ The associated display should provide the capability to distinguish significant levels of the system parameter controlled. ■ Feedback from the display should be apparent for any deliberate movement of a control.

Table A–6 Hardwired Panels—Controls (cont'd)

Component	When to use	Recommendations
Push buttons	■ Push buttons should be used when a control is needed for momentary contact or for actuating a locking circuit, particularly in high-frequency-of-use situations.	■ The push button surface should be concave to fit the finger. ■ The surface should provide a high degree of frictional resistance. ■ Large, hand or fist-operated, mushroom-shaped buttons should be used only as emergency stop controls. ■ With the exception of the above, all push buttons on a panel should have the same size and shape. ■ Each button should be labeled by a short and unambiguous text or graphic. ■ Button size should accommodate the largest label. ■ Labels should be consistent throughout applications. ■ Label should describe the result of pressing the button. ■ A positive indication of control activation should be provided (e.g., snap feel, audible click, integral light).
Toggle knobs (switch)	■ Toggle switches should be used where two (sometimes three) discrete control positions are required or where space limitations are severe.	■ When preventing accidental actuation is important, channel guards or lift-to-unlock switches should be used. ■ An open cover guard should not interfere with the operation of the protected device or adjacent controls.

Table A-7 Hardware Panels—Displays

Component	When to use	Recommendations
Meters (display devices): general		■ Pointer tips should be simple. ■ The pointer tip should extend to within about 1/16 inch of (but not overlap) the smallest graduation marks on the scale. ■ Pointer/background contrast and pointer size should be adequate to permit rapid recognition of pointer position. ■ Zone markings should be conspicuous and distinctively different for different zones. Use differently colored bands (color related to meaning) to indicate the normal operating range, limits, and danger range of a parameter. ■ Individual numerals on any type of fixed scale should be vertical. ■ Fixed scale with a moving pointer should be favored.
Circular meters		■ Scale values should increase with clockwise movement of the pointer.
Vertical meters		■ Scale values should increase with upward movement of the pointer.
Horizontal meters		■ Scale values should increase with pointer movement to the right.
Chart recorders	■ When a visual record of data is necessary; printers, provide exact numerical value and reference records.	■ The printed matter should be easily readable. ■ A minimum luminance contrast of more than 3.0 should be provided between the printed material and the background on which it is printed. ■ The printer should be provided with internal illumination if the printed matter is not legible in ambient light. ■ The print output should be free from character line mis-registration, character tilt, or smear.

Table A–8 Large Screen Displays

Component	When to use	Recommendations
LSD	■ To support or enhance the situational awareness of a group. ■ To provide an overview of the achievement of the main operating (and possibly safety) goals of the plant; it may also provide information on the overall plant configuration.	■ The most important step in designing the LSD is to determine which collective task(s) it will support; this should be done before designing the other aspects of the LSD. This will determine its content. ■ The conventions used for visual coding, labeling, etc., should be consistent with those used in the rest of the HMI. ■ If additional, user-specific information is required for an individual using the LSD, this other information should be presented on the individual's own display. ■ There should be an alternate means to provide the same information to the users in case the LSD becomes unavailable. ■ The minimum viewing distance for an LSD should be no less than half the width of the LSD. ■ The maximum distance should be determined based on the location of the users of the LSD when they have to carry out their expected tasks. ■ The legibility requirements (character height) can be determined using the same requirements (16 to 20 minutes of visual angle). ■ If the users of the LSD are not expected to be facing it directly (i.e., they may be "off-centered" with respect to the LSD), then the designer should ensure that the LSD remains legible in this condition. ■ Care should be taken that the line of sight between users and LSD is unobstructed. For projected LSDs, care must also be taken that there is no obstacle, physical or human, between the projector and the LSD.

Table A–9 Procedures

Characteristic or component	When to use	Recommendations
Document control	■ For every procedure	■ Provide version number on each copy. It is also useful to provide a date as this may serve as an additional cue to the operator as to how current the procedure is. ■ Put in place measures to ensure that obsolete versions are promptly removed from service and replaced with new versions.
Approval	■ For every procedure	■ Provide names and signatures of author(s), reviewer(s) and approver(s)
Individual steps	■ Within the body of the procedure	■ Write the procedure step as a bullet or numbered list with one instruction per line. Avoid using the narrative form. This makes the steps easier and faster to read. Background material (e.g., why this is that way) should be covered in a reference manual, in training, or both. ■ As much as practical, use proper grammar. This makes checking by commercial word processing software easier and, for most people, the resulting text will be easier to understand. ■ Use short sentences, preferably with one sentence per instruction. Again, this makes it easier to read and understand. ■ Use active voice. For example, say "Limit the speed to 120 m/s" rather than "Speed should be limited to 120 m/s." ■ Avoid negations, as the operator usually requires more time to understand them and runs a higher risk of making a mistake in doing so. ■ Use the imperative mode, as in a direct order. For example, say "Open the valve" rather than "Once the valve is opened…"

Appendix A—Design Guidelines and Components 163

Table A–9 Procedures (cont'd)

Characteristic or component	When to use	Recommendations
Recommended terms		■ Use the recommended terms, for example: <table><tr><th>Recommended</th><th>Avoid</th></tr><tr><td>IF</td><td>FOR EXAMPLE</td></tr><tr><td>THEN</td><td>EXCEPT</td></tr><tr><td>WHEN</td><td>BUT</td></tr><tr><td>AND</td><td>ONLY IF</td></tr><tr><td>OR</td><td></td></tr></table>
Conditional instruction		■ Conditional instructions should be written on separate lines. Also, capitalize and underline conditional terms. For example: WHEN the temperature reaches 120°C AND the flow is 66,000 l/hr THEN open the steam valve. OR WHEN the temperature reaches 120°C AND the pressure is 100 psig THEN open the steam valve.
Font		■ Use a font that is simple and a font size that is sufficiently large so as to be legible in the expected environment of use. Remember that procedures may be used in locations where the lighting is less than optimal. This usually requires that the font size be increased somewhat.
Calculation		■ Avoid having the user carry out calculations. If calculations are necessary, provide a job aid.

Glossary

Establishing a common language is often the first step to mastering a domain. This is especially true in the area of human-machine interface design. Below find a glossary of terms frequently used in this book.

Ergonomics, and/or human factors In Europe, the term *ergonomics* often covers both the physical and cognitive aspects of the interaction of people with machines. In North America, the term *human factors* is usually used for that purpose. In this book, we use the term "human factors engineering" to describe work aimed at improving a worker's health, safety, satisfaction, and productivity.

Function A process carried out by an agent (human or machine) to achieve a goal. A function can be broken down into sub-functions, sub-sub-functions, etc.

Human-Machine Interface (HMI) The means employed by a user to operate a machine, system, or process—for example, via a hardwired panel or computerized console. HMI also encompasses decision-support devices, such as operating procedures.

Mock-up Non-functional model of a system. For HMI design, mock-ups are normally used to test HMI concepts.

Prototype First version of a system, often with a limited functionality.

System — Interdependent staff, equipment, objects, and processes organized to achieve specific goals. A complete system includes installed equipment, materials, software, documentation, services and related support staff.

Tasks — A difference may exist between the operator's general task (e.g., producing paper on paper machine X) and his or her specific task, such as system start-up. For HMI design, the specific task is of particular importance. These tasks can often be represented, and understood, as a hierarchy of tasks, sub-tasks, sub-sub tasks, etc., until elementary actions are reached—for example, pressing a button. From this perspective, the operator's task includes the set of sub-tasks and elementary actions required to achieve specific goals.

Usability — The extent to which a product can be used by specified users to satisfactorily achieve specified goals effectively and efficiently in a specified context of use (ISO 9241-11:1998, definition 3.1).

Index

A

Accident situations 126
Alarm 42
    acknowledgement 101
    handling 53, 78, 81
    message lists 153
Alarm system
    management 18
    updating 83
Analysis document 38

B

Bar charts 151
Batch processes 96–97
Batch-level goals 98
Business case 88

C

Character height 8
Coherence 90
Color coding 9, 101, 143
    secondary coding 10
Commissioning 127
Common Industry Format 66
Communication 52
Compensating 43
Complementary evaluation 58
Computer-based systems 101
Computerized panel 102
    access to information 104
    limited awareness 103
    to reduce complexity 103
Context of use 19, 26
    analyzing 30–34
Continuous processes 97
Control center
    See Control room
Control room 111, 137
Controlling 43
Cost-risk-benefits analysis 91
Coupling 97

D

Data vs. information 104
DCSs *See* Distributed Control Systems
Decision-making 11

Decision-support system 136–139
   competence 137
   use for training 138
Design basis 89, 91
   documenting 91
Design solutions 20, 38–40, 49
   "reduce the ink" 46
   blending 109
Detailed design 45
Diagnosing 43
Dictionary of terms 100
Digital displays 152
Discount usability engineering 57
Discrete processes 96–97
Displays 161
Distributed Control Systems (DCSs) 1
   fault-tolerant systems 1
Documentation 38
Downflow 80

E

Egineering Equipment & Materials Users' Association (EEMUA 191)
   See also Alarm system management 18
Enhancing a system
   data collection 86
   data reduction and analysis 87
   design, evaluation, implementation, and follow-up 88
   prioritization and cost-risk-benefits analysis 88
   recommendations for improvements 88
   risk 88
   training benefits 92
Error prevention 48, 60
   mechanisms 13
Evaluation questionnaire 66
Exception handling 78

F

Font styles and sizes 101

G

General section 77–78, 80
Good practices 57, 58
Grafcet 60
Graphics and graphing 150

H

Hardwired panel 53, 67, 102–103
   reliability 103
Heterogeneous components 99
Heuristic evaluation 57–62
Hierarchical Task Analysis (HTA) 28–29, 45–46
Human error 13
Human-Centered Design Processes 19
Human-Machine Interfaces (HMIs) 1
   architecture 40–42, 45, 77, 89, 91, 98, 109
   composite 102
   design 7, 74, 86, 95
   design basis 84
   design process 19, 22, 77
   design work 73

Index 169

designer 99
evaluation 57
identifying information 105
information content 96
integrating 53
linking with system design 16
planning stage 19
requirements 73–74, 76
software-based 78
specification 52, 73–80
terminology 100
testing 109
Human-machine system 35, 137

I

Icons 143
IEEE 1233 75
 See also Human-Machine Interfaces (HMIs) requirements
IEEE 830 74, 75
 See also Human-Machine Interfaces (HMIs) requirements, specifications
Integration 99, 136
 decision-support systems 138
 strategies 101
Internal model
 See Mental model
ISO 13407 84

J

Joysticks 155–156

K

Knowledge-based behavior (KBB) 11, 13

L

Large screen display 105, 162
 guidelines 107
 technologies 106
Latent pathogen 14
Legacy system 89, 90
Legibility 8, 145
Long-term operating experience 92

M

Maintenance requirements 35
Management inputs 86
Management-by-awareness 15
Management-by-exception 15, 51
Mental model 12
Minimalist design 60
Mistakes 13
Mock-up 40, 52
Monitoring 43
 redundancy 44
Monitoring by wandering 103

N

Navigation map 89

O

Operating goal 49–51
Operating procedures 125, 136
  benefits 126
  computerized 132–134
  design principles and rules 130
  deviations 131
  event-based 136
  goals 131
  relation to HMI 130, 134
  relation to training 128
  symptom-based 136
  validation 129, 135
  writing 127–128, 130, 132
Operator complaints 85
Operator log 86
Operator monitoring 15, 95
Optimizing 43
Organizational requirements 20, 63
  specifying 34
Organizing a process 97

P

Permissible response delays 52
Piping and Instrumentation Diagrams (P&IDs) 39
Plans 31
Platforms 99
Presentation conventions 41
Process monitoring and control 99
Product quality 37
Production goals 50
Productivity 36

R

Rasmussen, Jens 11
Representative users 64
Results gap 84
  causes 84
Rule-based behavior (RBB) 11, 13

S

Safety 88
Safety culture 14
Safety goal 50
Safety-critical controls 103
Skill-based behavior (SBB) 11, 13
Skill-Rule-Knowledge (SRK) model 11
Slips 13
Soft controls 102
Software-based HMI 53
Specification document 52
Specification users 75, 79
Stand-alone units 1
Style guide 33, 138
Superficial complexity of panels 103
Supervisory Control and Data Acquisition System (SCADA) 1
Symbols 143
System custodian 86
System design cycle 16
System requirements specification (SyRS) 73
Systematization of operating practices 126

T

Tasks 31
Team performance 111
Touch screens 154
Training time 36
Training tools 126
Turnkey system 89

U

Units 101
Usability 3, 63
Usability testing 21, 58, 63–70
  disadvantages 63
User-Centered Design Process 24
  planning 24–26

V

Visibility 60–61
Visual coding 50, 53
Visual momentum 46
Visual pattern recognition 48
Visual perception 8
Visualization techniques 97

W

Waste 37
Wireless devices 107
  limits and advantages 108
  style guides 107
Working memory 10
Writer's guide 129